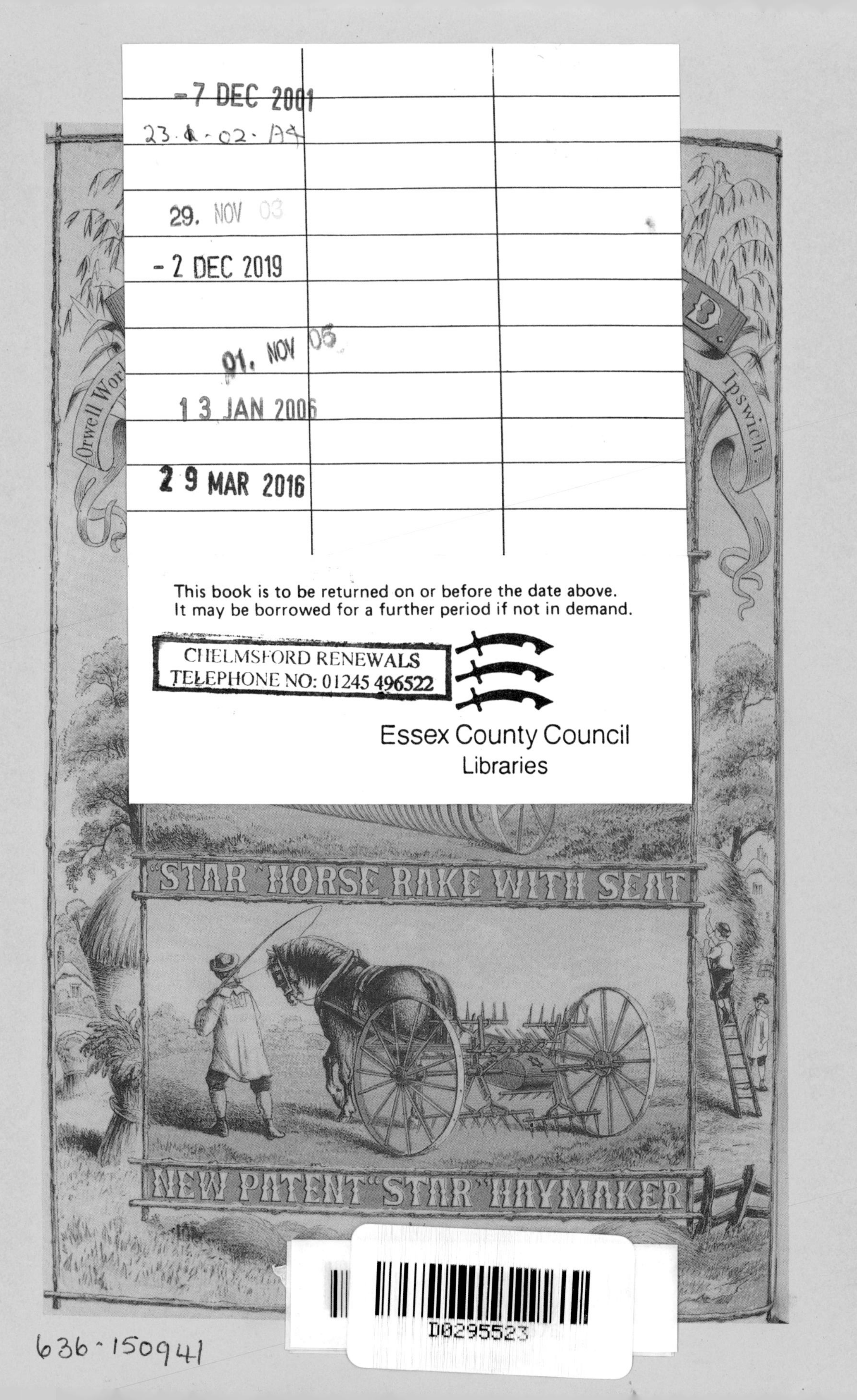
Orwell Works
Ipswich.
"STAR" HORSE RAKE WITH SEAT
NEW PATENT "STAR" HAYMAKER

THE HORSE IN HUSBANDRY

THE HORSE IN HUSBANDRY

JONATHAN BROWN

FARMING PRESS

First published 1991

British Library Cataloguing in Publication Data

Brown, Jonathan *1950–*
The horse in husbandry
1. Great Britain. Livestock. Horses, history
I. Title
636.1500941

ISBN 0-85236-217-X

Published by Farming Press Books
4 Friars Courtyard, 30–32 Princes Street
Ipswich IP1 1RJ, United Kingdom

Distributed in North America
by Diamond Farm Enterprises,
Box 537, Alexandria Bay, NY 13607, USA

Cover design by Mark Beesley
Phototypeset by Galleon Photosetting, Ipswich
Printed and bound in Great Britain by Butler & Tanner Ltd,
Frome and London

CONTENTS

A colour section appears following page 52

PREFACE

A PAIR of horses drawing a plough across a field is a fine sight; a fine sound, even, for their quietness contrasts with the noise of most of modern life. It is now, of course, a rare sight, to be seen only at working museums, shows and those few farms that keep some working horses. It seems but yesterday that it was a common sight, and yet it is already 35 to 40 years, in some places longer, since the demise of horse power on the farm.

The passing of the farm horse seemed to many to be the end of something that had lasted since time immemorial. In a sense that was so, for horses had been used in British farming for a thousand years or thereabouts. In another sense the age of horse-powered farming was remarkably short-lived, for it reached its peak with the agricultural revolution and mechanisation of arable farming of the late eighteenth and nineteenth centuries. By the 1920s the finely developed management of horse power on the land was already beginning to break down under the pressures of economy in labour and then of the introduction of tractors.

It is the horse-powered farming of the final era, from the late eighteenth century onwards, that is the subject of this book. The text and illustrations together attempt to show what went into the development of the final age of horse power on the land, and something of how that system of farming worked. Predominantly it is the arable and mixed-husbandry systems, such as sheep and corn, that are the main concern, as these were the types on which horse power was most concentrated. Pastoral farming had less need for horses. Inevitably, that means that, geographically, western regions are less well represented. Otherwise the coverage is the whole of England, Wales and Scotland. That is a fair area for a book of modest size. As a result, it is necessary to deal mainly in the general. Many of the details, especially the subtleties of regional practice, have to be obscured, although, that said, there were at least as many common strands as there were local differences in the management of horse-powered farming. However, this book does not seek to be exhaustive. There always remains the later, if not the last, word to be said.

JONATHAN BROWN
Reading
February 1991

ACKNOWLEDGEMENTS

THIS does not attempt to be a work of oral history. Reminiscences are a source, but they are of the sort that has been written down, and full reference to them is given in the references. However, I cannot fail to acknowledge the help of Bill Petch, who in a career of over 40 years worked on many farms in England. He read nearly all the text, set some things right and added vivid memory and anecdote. He has written much of his story down, and maybe in a future bibliography . . .

There are others to thank: my friends and colleagues at the Institute of Agricultural History, in particular Barbara Holden of the photograph library. My father prepared the typescript and drew the diagrams.

All the illustrations are courtesy of the Institute of Agricultural History and Museum of English Rural Life, University of Reading. Two particularly useful collections are those of the *Farmer and Stockbreeder* and of Eric Guy, a photographer whose eye for a good rural picture was of the finest.

THE HORSE IN HUSBANDRY

1

INTRODUCTION

IN January 1936 an article appeared in *Farmers' Weekly* under the headline 'There is a secure future for the heavy horse.' It was written by Edward Lousley, farm manager of the Lockinge estate in Berkshire. 'After considering the papers read at the recent mechanisation conference at Oxford,' he wrote, 'and from what I have seen in many parts of the country, I believe that for farm purposes horses will be needed for, say, the next ten years almost to the same extent as they are today.'[20]* Therefore, he thought, it was important for farmers to continue breeding Shire horses.

There was good reason for such an argument at that time. By 1936 the tractor had proved itself as a reliable and efficient machine, ideally suited for the heaviest work of the farm, such as ploughing, harvesting and heavy cultivation. But doubt remained as to the economics of the tractor for light work, for which it was overpowered, and for work involving continual starting and stopping. Again the horse still seemed the most economic power for small farms. Lousley demonstrated the efficacy of this argument by noting that on his estate 100 horses were still employed alongside the 11 tractors, and there were no plans to reduce that number of horses.

By 1946, however, the number of agricultural horses in Great Britain had fallen by more than 150,000, and the farm horse was clearly in irreversible decline. Two things had combined to make Edward Lousley's words seem hollow. The light, diesel-engined tractor on pneumatic tyres was a more economical proposition, even for the small farmer[50], and the Second World War had so speeded the course of mechanisation that by its end, almost all those farmers who had not already bought a tractor were probably considering it. Over the next ten years the number of horses in British farming dwindled further. The Ministry of Agriculture took its last regular census of them in 1958, effectively marking the end of the era of horse-powered farming in this country with a recording of only 84,000 horses.

The horse had been used in agriculture since at least the time of the Domesday Book. Then, however, it was very much junior to

* See page 107 for references.

the ox as a draught animal. Oxen did all the heavy work of ploughing, yoked together in large teams, most often of eight. Horses, on the other hand, were employed for the lighter work, such as harrowing. That remained the position at least until the mid-thirteenth century. During the later Middle Ages, as estate managers and farmers placed greater value on the speed and stamina of the horse, it began to displace the ox at the head of the plough. By the fifteenth century in East Anglia, which was the most progressive farming region of England, horses accounted for about 60 per cent of the draught animals on the large manorial demesnes.[30] Throughout most of the country oxen were still the predominant draught animal, but a trend in favour of the horse had set in, and during the course of the next 300 years the displacement of the ox was all but completed. By the mid-eighteenth century oxen retained a few strongholds: the more western and northern districts, such as northeast Scotland, and parts of southern England, such as the Sussex Weald. While there were still small farmers throughout the country who preferred oxen for drawing their ploughs, the horse had otherwise achieved supremacy.

From the late 1770s until about 1820, however, there was a revival in the use of draught oxen. Arthur Young reported in 1813, 'One of the most interesting circumstances I met with in Oxfordshire was the increasing attention paid to oxen as beasts of labour.'[58] He was pleased about that, as were most other leaders of the agricultural world at the time, who strongly favoured the employment of draught oxen. To support their case they put forward arguments about the strength of oxen, the ease of feeding and keeping them and the fact that an ox too old for the plough could still be sold to the butcher. These were familiar arguments rehearsed at least since the sixteenth century, and quite likely from

An illustration from the fourteenth-century Luttrell Psalter, showing a horse being employed in harrowing.

Horses pulling the plough. The title page of Edward Maxey's *New Instruction of Ploughing and Setting of Corn* (1601). The drawing is not the most accurate: one might suppose from it that the plough was harnessed to the horse by a chain from the end of its tail.

the time the first horse was harnessed to a plough. The only new feature in the debate was the suggestion that the exigencies of war with France meant that it was better to grow wheat for people than oats for horses.

The main effect of the wars was to restrict the supply of horses, as military demands encroached upon civilian needs, and in that way the wars acted as an inducement for farmers to employ oxen. Even with that and all the promotional efforts of Arthur Young and others, the revival of oxen was slight. Modernising agriculture needed the greater speed and efficiency of the horse, and thus, not only was the supremacy of the horse hardly challenged, but during the nineteenth century it was taken up increasingly in those areas where farmers had hitherto adhered to oxen for their draught animals. In the northeast of Scotland, for example, horses were more regularly used for the plough from the 1840s and 1850s onwards, at first in mixed teams alongside oxen, later on their own.[9] The conquest of such former outposts of oxen meant that, with but few exceptions, horse power was universal until the advent of the tractor.

2

THE RISE AND DECLINE OF HORSE POWER ON THE FARM

SINCE the eighteenth century the demand for power in farming has increased substantially. Agriculture has become more productive and more intensive. In order to achieve these results more energy has had to be invested in working the land. Until the First World War much of the additional power was supplied by horses. To them fell most of the work in the fields and carting, replacing slower, less efficient oxen and manual labour.

There were other sources of power: wind, water and especially steam. Steam engines found common use during the second half of the nineteenth century. They became almost the sole source of power for threshing and were used for barn work, where they drove crushing and grinding machines. Some farmers used traction engines for haulage. Some, too, employed steam ploughing and cultivating equipment, but that was a practical proposition only for farmers with large fields, in which the engines could manoeuvre in comfort. Important though these were, they were really just exceptions to the general reliance of farmers on horse power for arable cultivation.

The increased employment of horse power during the nineteenth century arose from the intensification of cultivation which required more work to be done on the land, and from the use of new implements and machines, almost all of which replaced manual labour with horse power.

The work of cultivating the land became greater as farmers took more trouble to prepare clean seed beds and then to tend the growing crops. In the eighteenth century the preparation of a seed bed usually involved ploughing the field once or twice, followed by a harrowing and a rolling. By the 1840s, however, farmers looking for high yields were no longer satisfied with so little work. Ploughing was commonly to greater depth than the usual 5–8 inches of the previous 50 years, and a first ploughing might be followed by a subsoil plough. Next, a heavy tined cultivator would be used to drag out the weeds, often two such cultivations being thought necessary. A second, and maybe a third, ploughing, followed by several harrowings and a heavy roller, ensued before the farmer was satisfied he had a seed bed

A two-horse cultivator at work, about 1910. This man, like many, evidently found it as comfortable to walk as to ride on the cast-iron seat.

for his wheat. These were the operations on light soil; stiff clays were likely to require even more cultivations.

More intensive rotations kept the fields in more constant cultivation. It was true that fallows had been ploughed, perhaps six or seven times during the season, to keep them clean, but the sowing of clovers, or especially roots in place of the fallows involved more intensive work on the land. For example, preparing the land for turnips on medium soils in Essex was reckoned to require three deep ploughings and three harrowings during the autumn. The land was then ridged up to lie over winter. In early spring

Three horses dragging a harrow over a hilly field in Kent in the 1930s.

Ploughing and sowing roots on ridges in Westmorland.

manure was spread on the field and ploughed into the furrows. Finally the land was rolled before the seed was sown.[2]

The turnip seed could be dibbled in by hand, but by the 1840s the seed drill was preferred 'as more expeditious and certain.'[2] Intensive cultivation was bringing with it mechanisation, and new implements drawn by horses in place of manual labour. The

The seed drill, from the *Illustrated London News*, 1886. Not only were three horses needed for this type of drill, sowing ten rows of corn, but, as they were harnessed in this fashion, with a pair ahead of the horse in shafts, two drivers were employed.

One of the best known of the new cultivating implements of the nineteenth century was Crosskill's clod crusher, introduced in 1841. It consisted of a series of serrated iron discs, of some weight. This illustration, used in the contemporary advertising, shows the three horses expected to be required.

seed drill, drawn by two or three horses, replaced hand broadcasting or dibbling. The use of the seed drill led on to fertiliser distributors and horse hoes. New all-iron harrows and clod crushing rollers which gave better results than the wooden-framed implements hitherto used also made their demands on horse power, for they often needed three horses to pull them. Some of the new tined scarifiers were also heavy. Many farmers used four horses for them, some as many as six.

During the second half of the nineteenth century the mechanisation of harvesting meant more new horse-drawn implements. On the corn field, the reaper, introduced from America in the 1850s, reached common use during the 1870s. It was superseded from the 1890s onwards by the binder, which tied the corn into sheaves as well as cutting it. Heavier than the reaper, the binder often needed three horses rather than two if the harvest was to be cut at reasonable speed. At the same time, the hay harvest was mechanised. Mowers, drawn by two horses, were introduced in the 1850s along with horse-drawn hay makers and tedders, and rakes. Hay sweeps, worked by two horses, one at each side, with a steersman riding behind, were introduced to gather up the hay for stacking towards the end of the century.

Horse engines, or horse gears, were another important source of power in farming. The use of animal-powered engines, which required a horse or other animal to walk round in a circle driving a gear wheel, was of ancient origin, but it was only applied to farming in the late eighteenth century to drive threshing machines. These early horse gears were large, with a main overhead gear 20 feet or more in diameter. They needed two or four horses to operate them and were usually in a special 'engine house' built on to the side of the barn in which the threshing machine was a permanent fixture. Numbers of these engine houses may still be seen, especially in southern Scotland and northeast England where farmers were in the vanguard in adopting the threshing machine.

Rees's *Cyclopedia*, published in 1819, illustrated a horse gear suitable, it was said, for portable threshing machines. It was, however, still quite large, with overhead gears, and a little cumbersome in appearance. Not until the 1840s did a more manageable and portable horse gear appear with the main gear at ground level. This gear was much smaller, about four feet in diameter, and engaged a small spur wheel which turned the drive shaft itself. More sophisticated versions had additional reduction gearing.

This was an important development, for as threshing was turned over to steam power, farmers soon discovered other operations for which the small horse gears were ideally suited. They were set to driving barn machinery: cake breakers, chaff cutters, root cutters, grinding mills, bean kibblers and the other equipment necessary for the efficient management of livestock. Some of the first dairy machines, such as cream separators, were driven by horse gears, and in the fields hay stackers and elevators were so powered. Nearly all of these horse gears were for one or two horse power and one horse was perfectly adequate for most purposes. There were, however, four-horse engines available for heavy work.

All these new horse-powered implements might be expected to result in more horses on the farm and the statistics suggest that that was in fact so. There were about 800,000 horses employed in British farming in 1812, according to the returns of the excise duty levied at that time. In 1870, when horses were included in the annual agricultural returns for the first time, 966,000 were recorded for Great Britain. By 1886 there were 982,000 and the following year, when mares kept for breeding were also included, the total stood at 1,034,000. The gradually rising trend continued until 1910, when the 1,137,000 horses then employed in agriculture represented the highest total ever recorded.

These statistics are misleading, though. The rise in the numbers of horses from the 1870s to the eve of the First World War coincide with a severe recession in agricultural prosperity. Arable farming in particular suffered; land was converted to pasture, some even left uncultivated for a while. These developments were not conducive to a growth in the numbers of farm horses, for pastoral and livestock farming needed rather fewer horses than arable. The statistics were inflated by the numbers of horses bred on farms for sale to customers in towns in need of horses for drays, delivery vans, taxi cabs and omnibuses, or for carriage and riding horses. These, really the farmers' stock in trade, were for various reasons included in the agricultural statistics. Certainly, most horses

The horse hoe was altogether lighter, and needed only the one horse. It was common to have a boy guide it along the straight path between rows. This is from the advertising for Garrett's horse hoe of the 1860s.

The reaper, which came into general use during the 1870s. These machines were all intended for two-horse traction, and were equipped with the single pole for harnessing. The horseman did not always ride the horse—most reapers had a cast iron seat to one side of the machine. This one was made by Samuelson's of Banbury.

were put to some work on the farm, but that was really to keep them exercised and give them experience of work. They were surplus to purely agricultural requirements. This was noticed in the 1850s in a report on the farming of Shropshire. There, many farmers were breeding horses for the towns of the Midlands, and this, it was said, 'explains, in a great measure, the unusual horse-strength observed on most farms. Four horses all at length in a plough; five horses in a cart, with perhaps 2 cubic yards of dung; and long useless teams in the waggons, are of frequent occurrance.'[52]

The increased demands on the horses to pull the new equipment were undoubted, yet despite that, the numbers of horses required in farming almost certainly changed little throughout the nineteenth century. They may even have fallen. Philip Pusey, writing in 1850, argued that numbers were being reduced, and he noted a drop of 74,000 between 1840 and 1848. He based this on the official statistics for horses claimed as exempt from excise duty because they were kept for agricultural use, and these figures, as has been seen, are not the most reliable of guides. Pusey nevertheless drew an interesting conclusion from them: 'I can see no other cause of the decrease than the reduction of plough-teams.' Two horses were drawing the plough where three or four had done before.[43]

Pusey here points to a most important influence, for ploughing held the key to the farmer's need for horse power. The peak demand for power was always for the autumn ploughing. Estimates made in the 1860s were that the horse power required in October was 40 per cent above the monthly average, and in November 38 per cent. A study in the 1920s also showed that

(*Facing page, top*) This alternative version of the reaper was, understandably, less popular, because it required an additional man. The bearded man holding the whip and reins was the driver. His companion, who had to rake the cut corn back behind the cutter bar, was probably employed as a day labourer, while the driver was one of the horsemen. The picture was taken in Suffolk in the early twentieth century.

(*Facing page, bottom*) A hay rake, from a catalogue for 1859.

(*Below*) Three horses to a plough in 1910, worked with a driver as well as the ploughman.

the peak demand for power was October, this time estimated at 25 per cent above the monthly average.[11] It followed that if the farmer had enough horses for his ploughing in October, he should also be sufficiently equipped for the other busy times of the year: the spring cultivations, haymaking and harvest. The introduction of horse-drawn seed drills, hay mowers and corn binders had the effect, then, of giving the horses fewer days off, rather than requiring the farmer to enlarge his stable.

Improvements in the design and manufacture of ploughs did have the effect of reducing the number of horses needed for a plough team. The heavy old types of plough common in the eighteenth century, with their big wooden beams and mouldboards, were expensive of energy in draught, most requiring three or four horses, even on light and medium soils, and as many as five or six on heavy clays. The horses were yoked in line, and commonly had a driver as well as a ploughman in charge.

At the beginning of the nineteenth century Arthur Young noted a farmer in Oxfordshire who had brought a Rotherham plough, and its ploughman, from Nottinghamshire, to try working with only two horses instead of the teams of three or four usual in his district.[58] The Rotherham was the first of a number of new types of lighter weight and draught that were produced during the eighteenth century. Joseph Foljambe of Rotherham patented designs for it in 1730, and later in the century James Arbuthnot, of Norfolk, and James Small, in Scotland, also designed ploughs that were calculated to require less draught to pull the share through the soil. From the late eighteenth century, improved techniques of ironfounding enabled the new firms of plough-makers, such as Ransome, Howard, and Hornsby, to produce strong and light ploughs with cast iron shares, breasts and coulters, and mouldboards of efficient design. All these new types of plough were intended for two-horse traction on medium and light soils.

(*Below and facing page*)
Two illustrations of ploughing from W. H. Pyne's *Microcosm* (1806) showing the large teams harnessed to some of the cumbersome types of plough used in the eighteenth century. A boy was employed to lead the horses. The team of four is shown with the woollen fringing around the housen which was then a common form of decoration.

At the end of the eighteenth century there were a few districts where their use was established. Among them were Norfolk, Suffolk and parts of the Scottish lowlands where Small's plough had achieved popularity—all areas in the forefront of agricultural progress. Elsewhere the smaller teams were employed only by the pioneers and experimenters, farmers like the one in Oxfordshire with his Rotherham plough.

By the 1840s the spread of the more efficient ploughs was making the plough-team of two horses common everywhere. A report from Cheshire in 1844 was typical of many: 'Ploughing with two horses abreast is now almost invariably the practice, where the iron plough is used.'[40] With the smaller team had come a change from harnessing the horses in line to harnessing abreast. Robert Bakewell was said to have been the first to harness abreast when demonstrating the efficiency of animals of his improved breeding in the eighteenth century. That may have been so, but certainly the use of only two horses to the plough commonly resulted in the adoption of yoking abreast. Saving the labour of the driver was the main attraction in making the change, for the ploughman now drove the team using reins. But it was also accepted as being more efficient. As another correspondent from Cheshire put it, 'the horses travel better together, are nearer to their work, and perform it more effectually.'[40]

Heavy soils continued to need more horses to the plough-team. The undrained clays in Gloucestershire were still being ploughed with teams of four or six in the 1840s, but in most other places three horses was becoming the normal team for heavy soils. Three in line were retained for wet soils in Cheshire; the same in the Fens in wet seasons, when the use of a team in line avoided compacting the soil. In Wiltshire three or four horses worked the ploughs on heavy land in the 1840s. The teams of three here were 'not worked at length, but invariably driven one before, and

(Above)
An iron plough drawn by two horses, guided by the one ploughman; north Wales in the 1940s.

(Facing page, top)
Three horses draw the plough up a hillside in Wiltshire in the 1940s, followed by the furrow press drawn by two horses in line.

(Facing page, bottom)
Two-furrow ploughs were often regarded as efficient and economical. They saved on horse power—three horses instead of four, with two single-furrow ploughs. They saved manpower, needing one ploughman instead of two. They had the disadvantage that it was not possible to make the ends of the furrows in a perfectly neat line along the headlands.

two abreast behind.'[32] In Essex, the heavy clays needed 'three powerful horses', which 'require the best description of food to enable them to perform their work.'[2] These clays of Essex became proverbial as 'three-horse lands', yet even that represented a reduction from the big teams of the eighteenth century.

Philip Pusey's estimates, therefore, were along the right lines. Farmers were becoming more efficient in their employment of horse power, at least in this important matter of the plough-team, and it was making a difference to the number of working horses on the farm. Arthur Young was quoting five or six horses to 100 acres as the number farmers needed. Estimates from the mid-nineteenth century onwards were that four to the 100 acres were kept, and that accords with the records of stables on many individual farms.

After the First World War the number of horses on farms fell markedly. The agricultural statistics for 1921 record 962,000 farm horses in Great Britain. Ten years later the number had fallen to 874,000, and by 1939, the total was down to 649,000.[1] At the same time the number of tractors was rising. The use of tractors had been modest before 1914, but during the First World War the government ordered 10,000 tractors as part of its ploughing-up policy. Expectations that this would lead to rapid mechanisation after the war, and the demise of horse power, proved unfounded. The use of tractors actually diminished slightly during the early 1920s, after which numbers rose slowly to 22,000 by 1930, and then more rapidly reaching 50,000 in 1939.

Once again, the statistics on their own are an imperfect guide. The process of mechanisation during this period was quite complicated, and the decline of the horse was in some respects

Four horses (the front left barely in view) are harnessed to this two-furrow balance plough.

Light carting tended to remain the horse's preserve into the 1940s. Spreading manure on the stubbles.

less rapid, in other ways more advanced, than the figures imply.

Non-agricultural demand for horse power was falling dramatically. Motor transport in the towns grew rapidly during and after the First World War, and the market for the horses bred for sale shrank. The breeding of draught horses collapsed to such an extent that by the mid-1930s farmers were complaining of a shortage of horse power. That was the burden of Edward

Two horses draw a wagon loaded with baled straw through a southern village in the 1940s.

Lousley's argument with which Chapter 1 opened. The demand for horses was still strong, he was saying, and there was no need for the breeders of heavy horses to give up their work. Similar views appeared elsewhere in the farming press of the time.

It was a strongly held view in the inter-war years that there were limits to the economical use of the tractor. The extra power of the tractor certainly paid for itself on all the heavy work of the farm, such as ploughing, drilling and cutting the corn, it was said. But for other work, such as harrowing and general light carting, the horse was reckoned to be the more economical. There were also several jobs on the farm for which it was claimed the horse was better suited than the tractor. The stopping and starting work in the harvest field as the stooks were loaded on to the wagons was one. The experienced horse that needed little driving held a definite advantage over the tractor, and on many farms it was not until the arrival of the pick-up baler that the horse lost this job. Some farmers retained horses for ploughing round the headlands, and in difficult corners of the field. A horse could turn a plough on a headland five yards wide, whereas a tractor in the 1930s needed nine or ten yards. A horse-drawn plough was often used, as well, to set all the ridges and to finish the stetches, while the tractor ploughed the rest of the field. That, of course, kept the horseman feeling indispensable.

Turning the ploughs at the headland.

(*Below*) Horses on the Lockinge estate in the 1930s. The addition of the hay loader at the back of the wagon was often held to require a tractor. Here a third horse was harnessed.

For reasons such as these, farmers and estate managers in the 1930s stressed the complementary nature of horses and tractors. Edward Lousley again pursued this line, noting that on the Lockinge estate about 100 horses were still at work and 11 tractors: 'one tractor to eight or ten horses is a very fair economical combination.' Even the most mechanised farms in the 1930s usually retained a few horses. The Westertown Farms in Aberdeenshire, for example, used three Fordson tractors to do most of the work on the 800 acre estate, but still had two horses for some duties.

However, there was more to the story, for horses were effectively being replaced by mechanisation in various ways. A stationary oil engine drove the grinding mill in the barn, and the horse gear could be dispensed with. Second-hand lorries were bought for transport around the farm, and to operate the hay sweeps. And the tractor itself offered enormous opportunities. While the Lockinge estate had bought eight tractors, 75 horses had been released. Having bought an expensive piece of capital equipment, farmers were inclined to use their tractors as often as possible, whether or not they kept their horses. A correspondent of the

(*Above*)
Horse meets tractor. This was commonly argued to be a good complementary arrangement, with tractor drawing the heavy seed drill, and horses on the harrow.

(*Left*)
But not always: sometimes the tractor had the harrow.

(*Facing page*) Harrowing on an autumn afternoon, Lancashire, 1939.

Farmers' Weekly noted that the convenience meant that horses might be left standing idle. 'I have seen a tractor hauling a self-binder while three or four able-bodied horses were running on a grass field no doubt thinking what a wonderful thing a tractor is!' he remarked.

By the mid-1930s the pressure towards mechanisation was becoming compelling. The running down of the nineteenth century's pattern of stable management, with its establishment of regular horsemen, each with his own team, was well under way. More horses on farms were effectively kept as pets, rarely put to a real day's work, than farmers would care to admit.

The Second World War drove harder and faster the urge to mechanise, and replacement of horses in farming became more thoroughgoing. There were four times as many tractors in 1946 as there had been in 1939. During the war, tractors came to provide more than a quarter of all draught power on the farm, and by 1951, Britain became the first country in the world to record fewer horses than tractors on its farms.

Government policies to promote increased production and efficiency encouraged these developments. New, light diesel-engined tractors and other technological developments such as smaller combine harvesters meant that what had hitherto seemed uneconomic for the small farmer was now perfectly feasible. Economists made detailed studies; others put the argument in more general terms. 'The food consumed by two horses will produce more than two hundred pounds worth of milk', claimed George Henderson, and for a small farmer that sum would keep a tractor in tyres and petrol for some time.[25]

Attitudes had been changing amongst farmers and their workers. Generations of countrymen who had been horse-minded were being replaced by those who were machine-minded, observed A. G. Street.[50] It was a change that had been at work at least since the First World War. Ralph Wightman thought that his generation, coming into farming between the two world wars, was perhaps the first really to dislike horses. They saw the people of the towns enjoying free time on Saturday afternoons and Sundays, and 'wanted something that didn't need to be groomed and fed when work was finished. . . . We were engaged in a war with father to make him buy a tractor.'[42]

As a result of these changed attitudes, farmers in the 1930s who had tractors were finding it easy to get men to drive them, while their horses were being rather neglected. John Stewart Collis found this regrettable, but an argument in favour of complete mechanisation: 'horses could now be released from their slavery.' A. G. Street, allying himself firmly with the machine-minded, agreed: 'being a horse-lover, I never want to see one wearing a collar again.'[20]

3

BREEDS AND BREEDING

THE modern breeds of heavy horse for agriculture were beginning to emerge during the second half of the eighteenth century. Before then there were only broad distinctions between different types of horse: packhorses and carthorses. Packhorses were used for transport and were usually light horses of small or medium size. There were numerous local types, principally from the southern and western counties of England and from Wales. The larger and heavier carthorses, used for draught work, derived from eastern and midland England, and from the central lowlands of Scotland, places where there was a stronger and older tradition of putting horses to the plough.

Most farmers undoubtedly employed carthorses of nondescript type, but there was in England a broad distinction between the black horses and the sorrels. By the mid-eighteenth century the sorrel, whose home was Suffolk, had already acquired a number of characteristic features. Arthur Young, recalling the old Suffolk horse of his youth, wrote, 'In some respects, an uglier horse could not be viewed; sorrel colour, very low in the fore-end, a large ill-shaped head, with slouching heavy ears, a great carcass and short legs, but short-backed, and more of the *punch* than the Leicestershire breeders will allow. These horses could only walk and draw; they could trot no better than a cow.'[59] The work of breeders during the late eighteenth century had effected several changes to produce a horse that Young could see as more handsome, lighter and more active, and yet still of great power and strength. It retained the solid barrel-like body, which gave the Suffolk its popular name, the Punch. The term sorrel to describe its colour was being dropped in favour of chestnut (chesnut to the true Suffolk man). By the end of the eighteenth century the Suffolk was clearly recognisable as a distinct and named breed. It was not, however, until 1877 that a breed society was formed, the Suffolk Stud-Book Association, later known as the Suffolk Horse Society, which published its first stud book in 1880.

When the compilers of the stud book for the Shire horse set to work in the late 1870s they, too, were able to trace lines of descent back to the mid-eighteenth century. But the emergence of this type as a distinct breed was far less clear. What became the Shire developed from the various strains of heavy horse bred throughout midland England, from the Fens in the east to Staffordshire in the west. Professor David Low, in his *Elements of Practical Agriculture* of 1834, noted the common characteristics of all these horses to be their large, sometimes excessive, size, broad breast, large and muscular thighs and forearms, and fairly short legs.

Colour varied greatly, from black through a range of bays and browns, and it was differences such as these that made establishing a common type out of local partisanship a difficult process. Not least of the difficulties was the name of the breed. David Low called it the Old English Black Horse, a formalisation of the common general descriptions of the horses as 'blacks', 'old blacks', 'strong blacks', and so on. There were those who preferred such terms as Lincolnshire blacks and Staffordshire browns. The breed society was founded as the English Cart Horse Society in 1878, an unsuccessful attempt at compromise. It was renamed the Shire Horse Society in 1883 after another old name, 'shire-bred', that had been quietly gaining popularity during the middle years of the nineteenth century.

The Old English Black Horse, a stallion said to be descended from one of Robert Bakewell's horses. Dishley Grange is the house portrayed in the background. The illustration is from David Low's *Breeds of the Domestic Animals of the British Isles* (1842).

The Clydesdale was the heavy horse of lowland Scotland, and, again, the origins of the modern breed can be traced back to the mid-eighteenth century. By the 1820s the Clydesdale was recognised under that name as an established breed in textbooks on Scottish agriculture. Although the original home of the horse was the plains of the Clyde valley in Lanarkshire, the breed, like the Shire, was based on many strains, including lines from Galloway, Ayrshire and Kintyre. Other centres of breeding were to the northeast, in Aberdeenshire, and to the south, in Cumberland.

The Clydesdale was similar to the Old English Black Horse, as it was usually black, with shades of brown and bay as the second main colour, and size, conformation and strength were also similar. David Low expressed the general view of his time that the Clydesdale was lighter, smaller and less strong, but that it drew 'steadily and is generally free from vice.' Breeders of Clydesdales were among the leaders in promoting heavy horses in the mid-nineteenth century. The breed society was established in 1877, and the stud book opened in 1878.[6; 33]

David Low listed a fourth type of agricultural horse, the Cleveland Bay. This breed originated in the Cleveland area of north Yorkshire, but was bred widely throughout the counties of Yorkshire and Durham. It might seem a surprising choice for inclusion in a list of agricultural horses, for this has always been classed as a light horse since it derived from the packhorses and the light general purpose horses of the region. The old breed of the eighteenth century was described as 'neither black nor blood'—free of any trace of carthorse or thoroughbred—and was a simple draught horse used almost entirely for agriculture.

By the time David Low wrote, the Cleveland Bay had 'a touch of good blood', the result of about thirty years' crossing with thoroughbreds to improve the breed's saleability as a riding and carriage horse. The Cleveland continued to be the mainstay of the agricultural workhorses of its area, but during the course of the nineteenth century was gradually supplanted, first by heavier local types of carthorse, then by Shires and Clydesdales. Even so, the Cleveland Bay continued to find agricultural employment into the 1930s, especially on farms where it was bred, for here the mares were usually put to some work.[14; 22]

The example of the Cleveland Bay serves to demonstrate the fact that, until the late nineteenth century at least, there was more variety in the types of horse used in farming than the simple list of the breeds of heavy horse suggests. Light horses of many types were used, more particularly in the western and upland regions, where arable farming was of less importance and consequently there was less need for great draught power. In Devon, at the beginning of the nineteenth century, the farmers had 'a small snug breed of horse, between the pack and the

The Cleveland Bay, another illustration from David Low, *Breeds of the Domestic Animals of the British Isles* (1842).

larger cart horses.' In northeast Scotland, the light and sturdy 'garron' was the native type which was worked alongside oxen into the mid-nineteenth century. The horses of upland Wales were generally of the cob and mountain pony types, together with some crosses with carthorses.[9; 53]

Heavy horses used outside the home territories of the main breeds varied considerably. They were often nondescript, mongrel and poorly bred: 'the horses generally used in the county of Gloucester are either all bad varieties of the old Lincolnshire blacks, or those of any other sort that can do nothing else.' Norfolk's horses represented a step of improvement on that. In 1858 it was reported, 'The Norfolk cart horses are hardy and useful animals, without being at all perfect in shape, or uniform in breed.'[3; 44] There were, in fact, some distinctive local types of carthorse. In Yorkshire there was a tall, black, fairly cleanlegged horse, which, while of coarse shape, was apparently of reasonable action. In Northumberland and Durham, there was the Vardy, which was a mixture of small local types and the big black horse.[10]

All these were of the lighter type of heavy horse, for through the middle decades of the nineteenth century, farmers were wary of the really heavy horses. The big black horses were what the draymen in the towns wanted, rather than what farmers needed. The black Lincolnshire carthorse, thought John Burke in 1844, was 'probably too slow for agricultural purposes, except on heavy, tenacious clay soils.'[7] W. C. Spooner, in a major article on the management of farm horses in 1848, declared that horses should not be too large, about 15 to 16 hands being a good height; 'they should be strong enough to render more than two horses in a plough a needless expense, and yet not so heavy as to impair their speed and activity.' For these reasons he preferred the 'many excellent compact cart-horses' to the massive black horse.[45] Similar views were shared widely throughout England. In the East Riding, for example, the farmers preferred quick-stepping horses, with which they expected to plough 1½ acres of their light soils in a day.[31]

This prevailing attitude in favour of the medium-sized cart-horse held back the advance of the black Shire type in English farming. These horses were almost entirely confined to farms in their homelands of the Fens and east Midlands, where there was a particular interest in the breed, or to farms breeding them for urban markets. Such farms were few, however, because profits in the mid-nineteenth century from the sale of the big black horses were small due to demand being restricted almost entirely to the brewers, and a few others who wanted showy horses. Even here some people found imported Flemish cross-breds a cheaper proposition.

Whether these were 'excellent compact cart-horses' must be left open, but these horses were clearly not of the pure heavy breeds. They were photographed in Sussex in the 1880s, harnessed to the traditional Wealden turnwrest plough—which some farmers still regarded as so heavy it really needed oxen to pull it.

Until the 1880s the profit from breeding horses of any type for sale to the towns was reckoned to be slight. In the 1840s John Burke put £26 as the cost of breeding a carthorse, from hiring the stallion to breaking in the colt at two and a half to three years old. The best price for the young horse sold at four or five years of age was likely to be about £40. A good dray horse of Shire type could fetch up to £60, but the costs of breeding it were far higher.[7] Returns from breeding and rearing horses for sale were thus reckoned to be far less than the profits from either cattle or sheep, and many farmers therefore gave up the business. This remained the position throughout the middle decades of the nineteenth century, into the 1870s, despite the fact that increasing demand from the towns pushed prices for draught horses up to £100 or more.[35] Farmers, then, were breeding mainly for their own use and remained content with the respectable, average carthorse.

Such considerations affected the development of the Suffolk and Clydesdale breeds to a far lesser extent. The Suffolk was a purely agricultural type, and met with more general approval among the leaders of the agricultural world. Suffolks were awarded most of the prizes for agricultural horses at the shows of the Royal Agricultural Society of England into the 1860s, including 14 first prizes.

The Clydesdale, too, was viewed with more favour by the leaders of farming opinion in England, because it was quicker in movement than the black horse. It had also a more ready sale to haulage contractors, railway companies and others beyond the narrow market of brewers and distillers. In Scotland and the northernmost parts of England the Clydesdale was becoming more widespread as a farm horse. It advanced steadily beyond the lowlands of its home territory to displace the garron and other types of light horse. There was also a trend towards the use of heavier horses in agriculture in England, where until the 1880s Spooner's 'compact cart-horses' had been used rather than the truly light types popular elsewhere in the country.

From the 1880s onwards the Shire horse came to greater prominence in English farming. Two coincidental developments furthered this. First, the founding of the breed society gave a boost, promoting raised standards of heavy horse breeding generally and of the Shire in particular. Second, farmers became more interested in breeding horses of the Shire type, because there was now more profit in them. While the depression that struck agriculture during the late 1870s resulted in reduced returns from most branches of farming, the demand, and prices, remained firm for horses sold to towns, especially for heavy draught horses. A report on the farming of Leicestershire and Rutland in 1896 showed that more carthorses than lighter breeds were being bred and that indeed, many farmers were giving up light horses because they could make no profit on them, whereas carthorses, especially Shires, were fetching good prices.[48] Shire breeding therefore spread to new areas, Yorkshire for example, where it had been virtually unknown before the 1880s. Since farmers who bred Shires for towns or export naturally used these horses on their farms and supplied their neighbours, horses of the Shire type steadily spread throughout English farming.

The Twentieth Century

By the twentieth century the three recognised native breeds of heavy horse were achieving dominance in British agriculture. If only a minority of the working horses on the farms were of pure breed, a large proportion had recognisably strong lines of Clydesdale or Shire blood in their crossing. Although the more general types of carthorse, and the light types, had not vanished from farming, they formed only a small part of the draught stock.

Pride of place amongst the heavy horses was now taken by the Shire. Its breeding was given prominence in the agricultural press, and its successes at shows were considerable. It was now being appreciated on the farms for some of the qualities that had been held against it in the 1840s and 1850s: its slow,

A champion Shire, Kirkland Black Friar, who was awarded first prize for Shire stallions at the Royal Agricultural Society of England's show held at Harrogate in 1929. Docking the tail of the horse had been a common practice since at least the late eighteenth century, but by this time it was purely a matter of fashion. In earlier years there had also been a good market for the horse hair.

steady pace, its docility and its massive strength. Farmers, and their workers, now rather liked these great horses with feet 'the size of dinner plates', with their strong muscular limbs. The pure Shires were tall, between 16 and 17 hands high, and weighed more than a ton. Preferred colours for the Shire remained black, bay and brown, but the cross-breds on many farms were likely to exhibit more variety. A prominent feature of the Shire was its feather, the hair round the feet and lower leg, which during the nineteenth and early twentieth centuries was a very profuse growth. More recent breeding has tended to favour cleaner legs, usually achieved by an infusion of Clydesdale blood.

The Clydesdale remained the dominant draught horse of Scottish farming. Few horses of other heavy breeds were used there, and the Clydesdale's only competition came from the mongrel and the light horses which continued to hold their place on some farms. As in David Low's day, the Clydesdale continued to bear many features that were broadly similar to the Shire. In some respects, indeed, the two types had perhaps grown more closely together, for during the mid-nineteenth century, breeders in Scotland had brought great numbers of mares from the English Midlands to improve the Clydesdale lines. The results of this traffic were to generate much heated, and ultimately inconclusive, argument as to the relative merits and demerits of

A Clydesdale mare. Brunstane Phyllis was champion mare at the Royal Highland Show held at Alloa in 1929.

Clydesdale and Shire, and even as to where the distinction between the breeds lay. It prompted the formation of the Clydesdale Society, part of the aim of which was to emphasise the distinctive nature of the breed.

In size the Clydesdale was about the equal of the Shire. The stallion stood at 17 hands, or a little taller, the mare at 16.2 hands. In conformation the Clydesdale was less massive, in both its body and its limbs, sacrificing strength to gain an action that was easier than the Shire's. As a result the Clydesdale was less ponderous, more agile and speedy in its movement. It was also noted for being high-spirited and mettled in temperament. 'The first yoking of a Clydesdale', noted Primrose McConnell, 'is usually a job for all the smartest and hardiest men on the farm to help at, and even then there is some fun.'[34]

The Suffolk was always regarded as the most purely agricultural of any type of heavy horse. Suffolks were steady; they were long-lived, with a good constitution, and possessed stamina that enabled them to work for hours at a stretch with little food, in comparison with the Clydesdale that needed copious amounts of food at regular intervals. Many regarded the Suffolk's lack of feather as a great advantage since it facilitated keeping the feet clean. They were small horses for a heavy breed, standing about 15.3 to 16.2 hands high, and, their smaller feet were regarded as an advantage for row-crop work. They were fairly slow, but with impressive strength and perseverance. The Suffolk, wrote William Youatt, 'would tug at a dead pull until he stopped.'[55]

Yet despite these advantages the Suffolk did not achieve widespread popularity as a farm horse, remaining virtually confined to

The Colonel, a Suffolk stallion, winner of first prize at the Royal Agricultural Society's show in 1862. It was owned then by Herman Biddell, of Playford, near Ipswich, a noted breeder of Suffolk horses, and also instigator of the founding of the Suffolk Horse Society and compiler of the first stud book.

Morston Earl, champion Suffolk at the Suffolk Show, Woodbridge, 1932.

its home territory of East Anglia, where it was dominant. Some Suffolks were kept in other parts of the country, such as the Lowther estate in Westmorland, where farmers or landowners had taken a particular fancy to the breed, as often as not because they liked the chestnut colour. But such examples were few to set against the advance of the Shire across England. The reason for the Suffolk's comparative lack of success seems to have been the fact that it was less popular as a horse for the towns than the Shire or Clydesdale. With a smaller market there was no great incentive for breeders outside East Anglia to take up the Suffolk. It was said that the blacks and browns of the Shire and Clydesdale were much preferred by purchasers in London and other towns to the chestnut of the Suffolk. It was argued, too, that the feet of the Suffolk were unsuited to hard urban streets. Certainly they were a point of weakness with the breed, being susceptible to ringbone and other complaints, which have been eradicated only during more recent breeding.[6] These problems were perhaps but symptoms of the fact that the Suffolk's success as an agricultural horse was based on conformation and characteristics that made it less suitable as a general draught horse than either the Shire or the Clydesdale. Nonetheless, a number of Suffolks did tread the streets of London.

Two imported breeds of heavy horse were introduced to British farming during the twentieth century. First was the Percheron, whose native home was the Perche district of Normandy. Although half-bred Percherons had been imported from America for some years during the late nineteenth century to pull London's buses, the first significant imports of pure-bred horses were made during the First World War. These were at the instigation of the military authorities, who had been sufficiently impressed with American Percherons bought for the army to want to investigate the pure breed. From these imports and others after the war, British breeding studs were established. The British Percheron Horse Society was founded in 1918, and the stud book commenced.

The use of the Percheron in Britain after the First World War was varied, but it was primarily an agricultural horse, especially on large estates in the eastern counties. On the Chivers farms at Histon, Cambridgeshire, for example, most of the stud of 150 horses was Percheron. Farmers appreciated the breed's stamina, soundness and hardiness, combined with its excellent temperament. The Percheron was of moderate size, between 16 and 16.3 hands high, compact, clean-legged, and with a head more refined in appearance than other heavy breeds. Pure-bred horses were either grey or black.

The Belgian horse was imported in large numbers during the years immediately before the Second World War. It was large, standing 16 to 18 hands high, with short, muscular legs and a massive, powerful body. Its colour was mainly dun or roan. Its temperament was placid, and it was a strong and willing draught horse, but, coming as it did towards the end of the era of horse-powered farming, the Belgian had but a small place in British agriculture.

Sir John, a Percheron stallion. 'With a view of improving the cart horses of the country this stallion has been imported and is now located near Borobridge in Yorkshire', reported the *Farmer's Magazine* in 1869. Such imports remained rare until the First World War.

Percherons at work in the Fens in 1941, hauling a modern wagon with pneumatic tyres along a track laid with concrete.

Belgian horses on Lord Radnor's estate, 1936.

Improving the Breeds

It is not entirely coincidental that the three native breeds of heavy horse can be traced back in their main lines of descent to about 1760, since it was then that new methods of breeding adopted during the eighteenth century were coming to bear results. The practice of gelding the majority of stallion colts, while retaining a small number of the best from which to breed by servicing large numbers of mares, was still a novelty in 1700, but was becoming general by 1750. The itinerant stallion, taken by his owner on a tour of the farms in a particular district, his impending arrival announced by handbills, had become a regular feature of country life.

These methods were an improvement on those that went before, but were nevertheless bound to be slow and haphazard as a means of raising the quality of breeds of horse. More effective for the establishment of pedigree lines was the practice of letting a stallion for a season to individuals or groups of farmers who had mares available for breeding. Robert Bakewell favoured this as an effective means of assessing the qualities of the stallions he had bred. Bakewell had taken the black horse of Leicestershire, crossed it with Dutch blood, and then applied his methods of in-breeding to establish improved lines. His work was much admired; Arthur Young thought Bakewell's stallions were the best he had ever seen of the Black type. Bakewell's work was influential. His example of letting his stallions helped establish this practice more widely, and by the early nineteenth century stallion hiring societies were becoming numerous.

By the end of the eighteenth century the standards of horse breeding were rising, and the quality of heavy horses was improving. The Napoleonic Wars and their aftermath, however, saw large numbers of horses sold off farms, many of them for export as buyers on the continent sought to make up the losses of the wars. With these sales went much of the best breeding stock, sold by farmers in need of money during a time of recession in agriculture,[10] resulting in a decline in quality which continued until the last quarter of the nineteenth century.

The return of more settled and profitable times to farming brought no improvement in breeding, since more money could be made from sheep and cattle than from horses. 'If farmers were nearly as particular about the individual qualities and pedigree, so to speak, of the horses they bred from, as many of them fortunately are in reference to the bulls and rams they use, they and the country would greatly benefit, and we should hear less of the unprofitableness and uncertainty of horse-breeding', concluded W. MacDonald in 1876.[35] Farmers were generally careless of the quality of both mares and stallions from which they bred. Only in Scotland, where Clydesdale

(*Facing page, top*) Not a type of horse immediately to be associated with draught work on farms, perhaps; still less in Hertfordshire. These Highland ponies provided the horse power for a small farm of 94 acres in Hertfordshire. The Land Army girls with the harrows show that the photograph was taken early in the Second World War.

(*Facing page, below*) An old Suffolk mare. This one was 20 years old when photographed in 1936, at work on light carting. Some horses, especially Suffolks, could live, and work, for several years more, but 20 was generally regarded as about the limit for a working life.

stallions were used in crossing, was there improvement in the quality of ordinary farm studs. In England it was left to a small number of enthusiasts to maintain the quality of the black carthorse, and to carry forward something of the legacy of Robert Bakewell's work. Nearly all of these were farmers in Lincolnshire, Leicestershire and Derbyshire, and they were responsible for breeding the horses upon which the main lines of modern Shire pedigree are based.

Breeding and Rearing Farm Horses

Writers on the subject of farm management were almost unanimous in the view that farmers should breed their own horses for work on the farm. The advantage of breeding on the farm, advised Henry Stephens, was that the young horses 'not only become naturalised to the products of its particular soil, and thrive the better upon them, but also become familiarised with every person and field upon it, and are broke into work without trouble or risk.'[47] W. MacDonald, writing in 1876, was equally certain: 'No practical man could deny that agricultural horses employed on the farm of their birth, or on which they were reared, are more healthy, hardy and durable, and altogether give infinitely greater satisfaction than those brought in after they are full-grown.'[35]

Many farmers ignored this advice, buying in their horses instead, whether as young horses from the breeder, as second-hand horses at farm sales or as older horses that had already spent one working life in the towns. Even those who did breed their own horses were likely to buy some, because they could not always maintain the fairly heavy investment which breeding entailed. The life of a working farm horse was usually between 16 and 20 years, which meant that an entire stable of 12 horses would need to be replaced over a period of 12 to 16 years, assuming that the horse worked from about the age of four. To achieve that, it was necessary to breed one foal each year, or ideally two or three to allow for the rather high mortality rate of infant horses from one cause or another. Maintaining such a continuous programme of breeding was more than some wished to undertake, even for their own use on the farm.

Farmers who did breed from their horses had to pay particular attention to the care of the mare and rearing of the foal. First they had to select a mare and hire a stallion. In the mid-nineteenth century, it was considered that mares should not be used for breeding until they were five years old. Likewise stallions, it was held, should not be used at an age of less than seven or eight. In the later years of the century these views were revised: 'under

Mare and two foals, photographed in Berkshire in the late 1930s.

the more liberal system of feeding which now more generally obtains [the mare] is frequently put to the stud at the age of two years, and, if well cared for, she does not suffer either in health or development.'[6] It was also accepted that stallions could be fit for service by the age of two or three, albeit to a limited extent for the first season.

Mares generally came on heat during the early spring, and as the gestation period was about eleven months, most foals were born between March and June. The convention which arose from the breed societies of declaring the age of a horse from 1 January, regardless of the month in which it was born, encouraged breeders to aim for early foaling, preferably January, so that their yearling colts might be at least 11 months old. Conversely it was a mistake to take early foaling so far that the foal was born in late December, for then it became a yearling after but a few days.

Mares were generally put to work throughout pregnancy. Heavy work in shafts was best avoided after about four months, but lighter work in chains could be continued, and a mature mare would work until the day she foaled. This not only gave the farmer the benefit of the mare's labour, but also prevented her from getting fat. The farmer also paid careful attention to the feeding of the mare, although there was not complete consensus as to what food should be given. A standard text of the early twentieth century declared that throughout gestation the mare should be 'liberally furnished with nourishing food in a concentrated form; bulky food of a low nutritive value is injurious.'[6] The Young Farmers' Club handbook of the 1940s tended to an opposite view, suggesting that grass and good hay, with small amounts of oats, were the most usual foods.

For the final days or weeks before the expected date of foaling the mare was stabled in a large, roomy loose box, well littered with straw. It was designed so that discreet watch could be kept so as to be ready for immediate action once the time of foaling arrived. After foaling, mare and foal remained in the loose box for a few days before being allowed out into pasture. They were kept away from other horses for a while longer, since they were likely to be sensitive of approach by other animals.

Weaning took from four to six months, during which time the foal was gradually introduced to grass and other more solid food. Six weeks after foaling a mare could return to light work. Opinion differed widely as to the advisability of this. Some held it to be beneficial on the grounds that on gentle work the foal could run alongside, which would be a useful first step in getting used to being handled by man. Others counselled against this because the mare would become hot, and this could upset her supply of milk. Mature mares, in practice, were given light work more commonly than not, simply out of the necessity for horse power on the farm. With young mares of two or three, the question of working them during weaning did not arise.

Once weaned, the foal was separated from its mother. At this point it ceased to be called a foal, becoming a colt if male or a filly if female. At a year old the young horse became known as a yearling, and at this age the colts not to be retained for breeding were castrated to be sold as geldings.

Breaking—training the horse for work—began in earnest usually at two years old. As a foal it had been made used to the halter, and as a yearling to some light leading exercises. Now, at two, the horse had to be made accustomed to a collar, it had to be shod, it had to learn to wear a bridle and bit, and it had to get used to being controlled by the reins. When those lessons had been learned, the horse could be taught its work. First it was trained to pull against long traces, and then gradually brought to work in chains. Next the young horse was introduced to work in harness, usually yoked with an experienced and docile horse. Finally it was brought to work in shafts, again yoked to an older animal. Early lessons often included taking a cart out on the road with three horses harnessed in line, the novice placed in the middle between two old horses.

Breaking the horse was important work which needed skill,

The first lesson for the foal was to be led by the halter. This one was presumably quite experienced when the photograph was taken, as it had evidently won a prize at a show, although where is not recorded.

care and sensitivity if the horse was to be willing and responsive. Care had to be taken with the horse's neck when the collar was first fitted, to make sure that it did not chafe; with the horse's foot when it was first shod; and with the horse's mouth, to keep it soft and thus easy to control. There were farms where the horseman who successfully completed the training of a young horse received a special payment. Conversely, the price of failure might be the loss of his job, since a horse that had not been broken properly was useless for work on the farm and was impossible to sell, except by underhand dealing.

By the age of four or five the horse should have been thoroughly trained and ready to work. It was then that it was sold, or that it joined the working stable of the farm.

4

WORKING WITH HORSES ON THE FARM

WHEREVER arable farming was of any importance, the stables acted as a focal point in the working life of the farm. It was in the stable yard that the farmer or his foreman was likely to hand out the orders for the day's work to the head horseman: so many teams to the plough, a trace pair to work a wagon, single horses for light work in the field, and so on.

The men who worked with the horses were usually specialists; the man at the head of the stables certainly was, although some of those who assisted him in the fields may have been ordinary labourers. These specialist horsemen were styled differently in various parts of the country. In the northeast of Scotland, for example, they were called simply horsemen, as they were in Essex and Suffolk. In Suffolk they might also be known as baiters.[16] In Kent they were usually referred to as wagoners. This was also the term in Yorkshire: hence the Waggoners Reserves raised by Sir Mark Sykes during the First World War. There were also wagoners in the east Midlands and Lincolnshire, although in the south of the county horseman was also used. Carter was another common term, used throughout most of southern England, and ploughman another widespread name. Less common was teamsman, the style used in Cheshire and Lancashire. In a few places an additional man was employed solely to feed the horses: 'carters groom but do not feed their horses' ran one mid-nineteenth century report from Gloucestershire.[39]

Whatever they were called, these horsemen were superior beings in the life of the farm. In the more purely arable regions they were the most important workers, and even where farming was mixed, the horseman's status was usually higher than that of other specialist workers, often higher even than that of the shepherd, although he was not necessarily the most highly paid. If there was a foreman, he, too, more often than not was appointed from the ranks of the horsemen.

Throughout much of the country, horsemen were among those, like cowmen and shepherds, who were employed by annual agreement. The hiring fair was the occasion for agreements to be struck, when the wagoner would stand in the market place with whip in hand, to betoken his work, and await the approach of prospective employers. In the more northern areas of England at least, this custom was still alive into the twentieth century.

The wages paid were not necessarily much higher than those for other labourers. Head wagoners in Lincolnshire, for example, were being hired in the 1880s for about £20 a year, which was actually less than the 13*s* a week paid to day labourers in that county. The wagoner, however, had more than adequate compensation since both his board and keep were also provided. In the north of England and in Scotland, the practice of having the men living in at the farm had still not entirely died out by the end of the nineteenth century.

The horseman also had the benefit of guaranteed employment, whereas the ordinary labourer might be stood down at slack times. That was the rule throughout the country even in those areas where hiring by the year was not usual.

When minimum wages were laid down during the First World War they included a differential in the weekly earnings of the horseman over the day labourer. In 1918 the rate for the horseman was 37*s* 10*d* a week, compared with 33*s* 7*d* for the ordinary labourers. This pattern continued into peacetime, although the rates of wages fell during the depression of the 1920s. The gap between the horseman's wage and the day labourer's was, however, effectively eroded by all the extra hours put in by the horseman at the stable, which were not paid as overtime. The horseman's advantages continued, then, to be in status and security of employment.

If the horsemen were the senior members of a farm's staff, there was also, on large farms at least, a hierarchy within the ranks of horsemen themselves. The head horseman was answerable to the farmer or his foreman, and under him came the second horseman, third horseman and so on down to the junior ploughboy. On large farms six or seven horsemen were not unusual, with two or three lads who generally did the odd jobs—light carting, taking horses down to the blacksmith—until they were deemed old enough and strong enough to take on ploughing.

Precedence was rigidly maintained, in a system found almost universally from the north of Scotland to East Anglia. The head horseman's authority over the stables was unquestioned. The farmer and foreman rarely interfered in the day-to-day running of the stables, and would often defer to the head horseman's judgement as work was being allocated to horses and men. The head horseman in his turn passed the farmer's orders on down the line of horsemen, and expected the respect due to his position. The head horseman always had the best horses, and he led in everything, however small. He was the first to enter the yard in the morning, first to feed his horses, first to harness them. When the teams left the yard to go out to the fields the head man led, followed by the others in strict order. At the end of the day it was not politic for one of those on the lower rungs to reach the yard before his seniors, even if he had been working in a different field.

During the twentieth century this rigid hierarchy broke down. The First World War provided a strong impetus. The loss of farm labour to the forces was particularly heavy from amongst the horsemen. The army purchased almost all the good horses from the farms, and some horsemen, finding that working with the poor quality horses that were left was too much to bear, enlisted themselves. There was, apart from such individual circumstances, a high proportion of horsemen of military age, whereas ordinary labourers tended to be older. As a result more unskilled workers had to be given work with the horses, leading

(*Right*)
This was the style of riding favoured by horsemen almost everywhere. Rarely did they sit astride the horse's back. It was the usual practice to lead or to ride on the nearside (left hand).

(*Facing page*)
The top photograph shows a carter and his horses, Wincanton, Somerset, 1900, evidently just back from the fields, the horses still to be unharnessed and groomed. Both are light horses, as was not uncommon in this part of England.

In contrast, the horses in the bottom photograph, which appeared in *Country Life* in 1911, have been groomed to spotless condition to pose with their horseman.

to farmers' complaints that men ran away from the horses when any difficulty arose, and that 'mere lads' were being entrusted with the carting.[4; 15]

During the years of depression in agriculture after the war, farmers forgot those complaints and continued to employ unskilled men with the horses in order to save on wages. Thus men coming into farm work during the 1920s and 1930s were likely to find themselves employed in all work, and be told to take a horse out in a wagon, without any experience of harnessing it. Where a horseman was employed to have charge of the stable, he often had to entrust much of the carting and field work with horses to day labourers. Many farms, especially if the number of horses kept was small, simply had no horseman as such. One of the general labourers was responsible for the stable. Even so, on large farms and estates, where a dozen or more horses were stabled, the old spirit lingered down to the Second World War.

Managing the Horses

Usually each horseman had his own horses to look after and work with, either a pair or a team of four. It was a major advance in the young worker's career when he moved from ploughboy to having a team of his own.

This arrangement of man and team was universally regarded as beneficial, enabling horse and man to become used to each other. 'When the peculiar tempers of each party are mutually understood, work becomes more easy to both, and more attention is bestowed upon it', wrote Henry Stephens.[47]

There could be the drawback of inflexibility, however. W. J. Malden, writing in 1896, thought that this was a major barrier to efficiency. 'It is one of the great objections', he argued, 'to the system which provides that a man shall work with and attend to a particular team [when] two weak horses . . . get sent to a heavy day's ploughing, while the strong ones are set to work light harrows. It is the farmer's place to arrange what work he will have done, to see that the horses best suited to it are put to it, and that the man or lad sent with them is suitable for the work.'[38]

Although the horsemen jealously guarded their teams, there was some flexibility. This was essential to make up the teams of four for wagons and three for the binders.

There was flexibility, too, arising from the fact that neither men nor horses were permanent fixtures. The men might move on to another farm, and horses were sold or died. Horses were also put in the charge of different men from time to time. It was usual for the young lads to have the oldest horse, or the horse kept for

Teams of Suffolks setting out for work on an estate in Buckinghamshire in the 1930s, with the head horseman in the lead.

(*Below*) Elevenses: the plough team takes a break for a meal from the nosebags. The bags were usually made of cocoa matting or manilla hemp.

odd jobs (quite often the oldest was the odd-job horse). Although the head man preferred to have the best horses, he was the most skilled and experienced, and thus would often take young horses in hand, passing on one of his old team to another man.

The system of working with regular teams was also reckoned to foster high standards of care. 'Upon the whole', wrote Henry Stephens again, 'there exists a good understanding in this country between the ploughman and his horses; and, independently of this, few masters are disposed to allow their horses, to be ill treated, and there is no occasion for it.'[47] The Young Farmers' Club's writers went further: 'No one, not even that proud man, the shepherd, takes more pride in his animals than a good wagoner.'[21]

That on the whole was true, but the world of farm horsemen had its share of the impatient and the short-tempered, who would resort to sticks or sometimes more painful instruments at the least sign of uncooperativeness from the horse. There were some who would have a nail or spike protruding from the traces of a ploughhorse to keep its partner from walking too close, since a sharp spike in its flanks would certainly keep the errant horse in its proper track. The horse that had been badly broken in could find the rest of its life miserable, as its hard mouth made it less responsive, and in consequence a trial to the impatient horseman. Apart from beating the horse when it failed to respond, some resorted to dubious practices to make the mouth more supple. These even included rubbing the mouth with sublimate of mercury, which was poisonous.

Again, as farmers cut back on labour after the First World War, the system of a horseman caring for and working with his own team began to break down. Young men with little or no experience of animals were hired and often expected to take on general work with any horses available. On some farms the horseman's main job was little more than managing the stable to supply horses for day men to use.

The Working Day

It could be hard for the farmer who moved from one part of the country to another. Primrose McConnell was brought up on a farm in Ayrshire. 'In my youth', he wrote, 'when I served on my father's farm, I worked in the field for ten hours daily, finishing at six p.m.' When he moved south to Essex at the end of the 1880s he found many differences in practice, but none amazed him so much as the horsemen's working day, for they finished their work in the fields at 2 pm. McConnell could not get used to that: 'the two o'clock idea sticks in my gizzard, and is likely to choke me.'[34]

There were two different patterns for the day's work with horses. The 'one-yoke shift' involved continuous work for seven to eight hours, with only brief rests, the shift ending early in the afternoon. The 'two-yoke shift' comprised two sessions, or 'half yokings', morning and afternoon, each of four or five hours. Between sessions there was a two-hour break for dinner, and work finished between 4 and 6 pm.

Apart from the shock of encountering a system so different from that to which he was accustomed, Primrose McConnell's distaste for the one-yoke shift was based on the apparent inefficiency and wastefulness of 'stopping in what, to me, is the middle of the day.' Even so, he had to admit that the amount of work done by his ploughmen in Essex was little different from Scottish practice. Fellow Scotsman Henry Stephens, in his *The Book of the Farm*, published in 1849, had also found working in one yoking 'highly objectionable.' To work for so long without a full break must be bad for the horses, he thought. 'Common sense tells a man that it is much better for a horse to be worked a few hours smartly, and have his hunger satisfied before feeling fatigue, when he will again be able to work with spirit, than to be worked the entire number of hours of the day without feeding.'[47] The Young Farmers' Club handbook, *Farm Horses*, took a measured approach: 'there is something to be said for both.'

Of the two practices, the two-yoke shift was the more common. Universally so, it seemed to Primrose McConnell, who mused as to why his corner of Essex should be out of step with the rest of the world. Certainly Essex and Suffolk were strongholds of the one-yoke system, and George Ewart Evans has argued cogently that this was a legacy from the days of ploughing with oxen. The time required to yoke and unyoke a team of eight oxen was so long that it was found better to work in one stretch and finish early, rather than break in the middle of the day. Even the terminology of yokings, and one- and two-yoke shifts, derived, Evans notes, from the days of oxen.[16]

What is unanswered is why East Anglia, where oxen were replaced by horses at an early date, should have maintained the one-yoke shift, whereas in areas such as northern Scotland, where oxen were still employed into the nineteenth century, the two-yoke was apparently universal. Perhaps the availability of a horse like the Suffolk, which had the capacity in its belly to sustain long hours of work with little rest, had something to do with it. And there was the usual straightforward contentment with established practice. Primrose McConnell found the farmers of Essex not in the least interested in changing their system.

However the day was organised, it started early. It was common to lead the horses out to the fields, between 6 and 6:30 in the morning, although some started as late as 7, especially in winter.

Before that the horses had to be fed, groomed and harnessed, for which at least an hour was needed. Most horsemen, therefore, were at work by about 5 o'clock. If they had not breakfasted before work, half an hour was allowed, but if the horse was difficult to catch and bring in from the field, or there were problems with the feed or harness, then the horseman's meal had to fit in as best it could.

At the other end of the day there was a similar amount of work in the stable yard. The horses had to be unyoked, groomed and fed again. Even those working the one-yoke shift would not finish before 5 pm at the earliest. On the two-yoke system a working day of ten hours was often achieved, the teams not leaving the fields until 6 pm. In winter, however, 4 o'clock was equally as common a finishing time. The men, therefore, finished in the yard between 5 and 7 pm. However, final feed was not until 8 or 9 o'clock, when the men had to give the horses their supper and bed them down. In the north of England and in Scotland all the men returned to their stable, whereas in southern counties, it was usual for just the head horseman to do the job. In the 1890s, at least one Scottish farmer reported that the 8 pm feed had been discontinued on his farm, but 'more out of regard to the convenience of the men than to any known advantage to

Coming home: the team leading the plough back to the yard. The ploughman has his coat hung on the hames of the horse on his left. The amount of decoration worn suggests that they had been to a ploughing match.

the health of the horses.' Instead, one of the ploughmen looked in briefly to see that all was well in the stable.[36]

Timekeeping could be strict. The head horseman would expect to lead his teams out of the yard on the dot of starting time on the farm. In the same way, meal breaks were taken at the prescribed times, and work stopped at precisely the official hour.

Stabling

'When it is remembered that, in winter time, farm horses often spend sixteen hours out of the twenty-four in the stable, the importance of a healthy dwelling house is at once seen to be very great', wrote W. Fream in his *Elements of Agriculture.* The stable, he went on, should be 'commodious, freely ventilated, well-drained and withal warm and comfortable.' The horse in a warm stable would require less food to keep up its bodily temperature than the horse in a cold, draughty stable. 'At the same time, it is a serious error to secure warmth at the sacrifice of pure air.'[23]

Fream perhaps had in mind the practice, which had for long been common, of attempting to ensure warmth at all costs. John Burke, writing in the *Journal of the Royal Agricultural Society of England* in 1844, had some sharp comments to make. 'The plan of stopping up every aperture at night and excluding the air is one which is very generally adopted by great numbers of people, alike ignorant of the injury they thus inflict upon their horses.'[7] There was plentiful evidence that an airless stable could cause all manner of disease, from grease and mange to sore throats, influenza and blindness.

Nonetheless many farmers and head horsemen persisted in the belief that only by coddling the horses in extreme warmth would they be kept in condition. In particular, it was widely held that the animals' coats were improved and kept glossy if the stable was free of draught and 'as hot as a conservatory.' Holes were, therefore, blocked up with straw and the bottoms of the doors sealed with sacks. In stables with few windows and a low ceiling, a lot of heat could build up. 'First thing in a morning the air was fit to knock you down and choke you on opening the stable', an old Yorkshireman recalled to J. Fairfax-Blakeborough, 'One wonders how horses' eyes, feet and lungs stood it.'[18]

The prevalence of such views meant that old stables often were small and cramped. They were no more than 16 feet wide, and the individual stalls 5 feet wide. Ceilings were low, no more than 8 feet high, both because of the desire for warmth and cosiness, and because it was common practice to have a hay loft above the stable, allowing hay to be conveniently dropped down to the racks above each stall.

The farmer's stable, from a painting of the 1790s by George Morland. It is difficult to know how accurate a depiction this is, but none of the neatness and order advocated by the writers of nineteenth century farming manuals is apparent.

All writers on the management of horses from the 1840s onwards hammered home the message that stables should give more freedom to the horse. There should be more room, with the width of the stable about 20 feet, the width of the stalls at least 6 feet, and height to the ceiling 10 feet or so. All agreed that the ceiling should be open to the ridge, with no hay loft. Some were unhappy with the hay racks in the stalls, arguing that they created too much dust which could harm the horse's eyes. Floors, it was felt, should be durable, even and easily cleaned, of concrete or closely laid setts, and there should be ventilation and light. The process of conversion to these ideas was, however, a slow one.

At quite the opposite extreme from the desire to keep the horses warm and snug was the practice of keeping horses out of the stable for as long as possible. During the summer they were pastured on grass. This was a matter of 'a few weeks' according to Henry Stephens, and in Scotland, where the season is shorter, midsummer to harvest was a common reckoning for the time horses could be left out. Further south the period was often extended as long as possible, beginning earlier in the spring and not ending until well after harvest. In Shropshire, in the 1850s, farmers keen to breed good, hardy work horses reportedly left their animals out until December, and some of the young ones even into January.[52]

Keeping the horses in the field was not always popular with some of the men. In the autumn they 'haggled the foreman until he allowed us to sleep the horses in the stable', recalled one old horseman.[13] Even in the dark of winter, work in the stable was often preferred to trudging across the field to entice the horses into the yard. It was time-consuming, even with willing horses.

In some areas, horses were turned into a straw yard for the night throughout the winter. Arthur Young, at the beginning of the nineteenth century, reported that a growing number of farmers in, for example, Lincolnshire and Oxfordshire, regarded this as a better way to keep their animals healthy than keeping them in the stable. Walls provided shelter, there was good deep bedding and hay in the racks, and there was an open shed if it rained.[56; 58] This was the practice still followed in Essex when Primrose McConnell moved there. 'Such a system', he thought, 'could only have originated in a district where straw was cheap and plentiful.'[34] Indeed, many farmers in the eastern counties of England kept their horses in yards all year, never turning them out to pasture. They argued that they saw no virtue in the horses competing for grass, and in these arable districts, where pasture was not plentiful, they had a point.

Turning the horses out into a yard had the great advantage that there was no need for mucking out the stable, although the yard had to be kept well littered with clean straw. When horses were kept indoors, cleaning the stables was one of the regular jobs for the morning, and one which seems often to have been neglected. John Burke in 1844 did not mince his words. 'Cleanliness in the stable', he declared, 'is . . . a point to be strictly attended to, on every occasion, and one to which for the most part the farmer is lamentably inattentive. There is scarcely one farm in fifty that is properly and thoroughly cleaned out every week.' As a

A yard for the young horses. A farm in Cambridgeshire during the Second World War.

result horses spent too much time standing on wet litter, with consequent risk of disease to their feet. Laziness on the part of the horsemen was at the root of this state of affairs, according to Burke. The pressure of trying to fit all the tasks into a busy morning's schedule would be the reason cited by many horsemen, certainly after the First World War. Henry Stephens offered another explanation, saying that many old stables were too cramped and dark for even the best of workers to clean properly.[7; 47]

Feeding

It was a maxim of the management of horses that the limited capacity of their stomachs meant they needed regular meals and should not be expected to work for long hours without food. Members of the Womens Land Army in the Second World War who enrolled on the correspondence course in farm management had this fact stated simply in the study notes: 'The working horse should have three meals a day.' Those, naturally enough, were breakfast, mid-day dinner and evening, but times in practice varied considerably. Some farmers also fitted in a fourth meal, at least during the busiest seasons, if not all year. In Scotland, where the day's yokings could be ten or twelve hours in total, it was common to have a tea-time meal at about 6 pm, and another evening meal at 8 pm. By contrast, the one-yoke day as practised in East Anglia tended to compress the daily meals. After a good breakfast the horses worked until mid-morning elevenses, which consisted of a nosebag meal taken in the fields. The main meal of the day was in the afternoon between about 3:30 and 5, after the teams had returned to the yard.

Horses needed ample time to digest before starting work. The mid-day break on the two-yoke system was thus generous, commonly from about 11 until 1 o'clock, allowing good time for a meal and the walk from the fields to the yard and back again. When work was in the farthest fields, this meal might be from nosebags. Henry Stephens, for one, heartily disapproved of that. 'This plan may do for a day or two in good weather, on a particular occasion', he declared, 'but it is by no means a good one for the horses, as no mode gives them a chill more readily than to cause them to stand on a head-ridge for even half an hour in a winter day, after working some hours. A smart walk home can do them no harm.'[47]

Eating between meals was also allowed. During the summer the horses had pasture in the evenings. In winter there was hay in the stable. During the day there were short breaks when they might have a nosebag.

Food for the horses was known as bait—hence the term baiter

▲In the early twentieth century, the cheaper three-colour process made coloured postcards based on photographs popular. Several series depicting rural and farming scenes were published. The card above shows adjustment of the harness on the team harrowing.

▼Ploughing in the orchard, followed by sowing by hand.

▲ A ploughing scene in upland Britain.

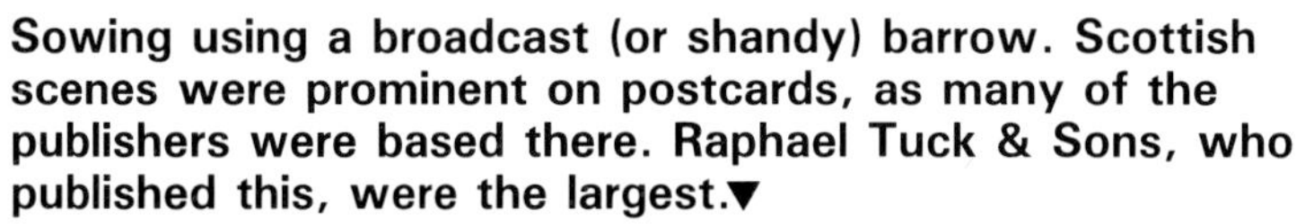

Sowing using a broadcast (or shandy) barrow. Scottish scenes were prominent on postcards, as many of the publishers were based there. Raphael Tuck & Sons, who published this, were the largest.▼

Haymaking in Hampshire. A well-matched pair of horses with a mowing ◀machine.

Two horses take a heavily loaded cart of hay carefully through the field gate.▼

◄*Noon*, from a print by J. Harris of a painting by J. F. Herring (1795–1865). Herring's main activity was painting racehorses, so it is perhaps not surprising that the three horses pulling the plough have more of the thoroughbred about them than was usual in farm horses.

▲Cutting oats with a manual delivery reaper.

Two binders at work in Scotland: one of Tuck's 'From Seed to ▼Grain' series.

▲ In upland areas of Britain, low, flat-bodied carriages were used for carrying hay. They were known variously as sledges or wains, and were used in preference to carts. Wagons were virtually unknown in these areas.

▼Loading the grain on a cart, from Tuck's 'Farm Life' series.

▲ Shoeing the grey mare. The peaked collar and the style of cart saddle favoured in Scotland are clearly illustrated. Another of Tuck's cards, this example was posted in 1906.

One of the means of selling horses bred on the farms to urban users was through horse fairs. Alfreton, in the Shire-breeding ▼ county of Derbyshire, had a horse fair of some local note.

for the horseman—while the wicker sieve in which food was served was known as a baiting sieve and the stable was sometimes called the bait stable. The bait was given, for the most part, as a mixture of oats and chop. Oats provided the concentrated food, highly nutritious in a small bulk and were usually crushed, since horses have difficulty digesting whole oats. Chop was clover hay or oat straw chopped (chaffed) into short lengths. This was a bulky food, less nutritious, but important for the mixture because it needed to be chewed properly, and this in turn helped the horse digest the oats.

Oats were always the staple diet of the horse. They were the best concentrated food, although as the Young Farmers' Club handbook noted, this was for 'some reason not properly understood.' The quantity given varied widely from farm to farm. There were several reasons: the nature of the work the horses were doing; the type of horse (the big Clydesdales and Shires needed more than the average or light draught horse); and local views as to the value of different foods.

According to A. G. Street, the average working horse consumed three bushels of oats a week, or nearly 40 sacks in a year (1½ cwt to a sack of oats).[49] There was some polemical exaggeration here, for Street wished to demonstrate the extensive acreage of farmland needed merely to feed the horses. Recorded regimes of feeding rarely approach Street's figure. Surveys of practice in the nineteenth century reveal that while some farmers allowed their horses no more than one bushel (42 lb) of oats per week, most fed one and a half to two bushels (63–84 lb), with very few actually giving more than two bushels a week.

At the beginning of the twentieth century, 12 lb of oats a day was recommended, and the Young Farmers' Club booklet published during the Second World War followed that line. Assuming the horse was to have the same rations on Sunday as the rest of the week, and again practice in this varied considerably, then the feed of oats amounted to two bushels a week. All these figures were for the horse in normal work, and at times of lighter work, less concentrated food might be needed.

Other concentrated foods were fed in addition to or as substitute for part of the oats ration. The most common was beans. There were farmers who thought these were at least as valuable as oats, and some in the mid-nineteenth century fed their horses only a bushel of oats a week with perhaps twice as much in beans. At the beginning of the twentieth century, rations, recommended as being more scientifically based, included about 3 lb of beans to go with the 12 lb of oats each day. Maize and bran were also commonly fed. Barley in its various forms, including sprouted and malted barley, always had its advocates, and gained support from numbers of veterinary scientists who vouched for its value.

Few farmers, however, fed it to their horses. Among the exceptions were farmers in Norfolk, where barley was an important crop, and, more generally, farmers with access to breweries who found brewers' grains to be a useful and inexpensive food.

Whatever the rations of corn, farmers intended them to be administered strictly; the horses should have enough to be well fed, but no more, in the interests both of economy, and of keeping the horses in good condition. There was a corn bin in the stables, and into it was measured the corn sufficient for the day, two days or the week, depending on the capacity of the bin. The bin was kept locked, and only one man had a key—the head horseman, the foreman or maybe the farmer himself. In this way control was maintained over the rations.

The horsemen, meanwhile, had a straightforward view that their horses were entitled to unlimited supplies of food of all sorts. Not unnaturally, they interpreted any restriction on supply as meanness, and set about thwarting the farmer's intentions as best they could. Many tales are told of how horsemen purloined extra supplies of corn. They would succeed in getting hold of the foreman's keys to the corn bin or the granary for long enough to have a copy made. Where granaries were built on staddlestones the horsemen used to crawl underneath and drill holes through the floor. They would even make off with a sack or two of corn while threshing was in full swing, and hide them for future use.[29]

There were other bulky foods besides chaffed hay and straw. Root vegetables were often added to the chop. Again, practice varied widely, both as to the choice of food and the quantity. Turnips were the most commonly given, fed during winter and spring. In the mid-nineteenth century some farmers, especially in Scotland and the north of England, gave their horses quite large quantities of turnips—280 lb, even up to 500 lb a week, compared with 112 lb or less which was the more common quantity. In some southern districts such as parts of Gloucestershire, carrots were favoured. Arthur Young noted that farmers in the Sandlings district of Suffolk had taken to feeding their horses with the carrots that could not be sold as table vegetables, a practice they continued through the nineteenth century.[39; 59] Potatoes were another possible choice, and swedes were often recommended, although the horse itself apparently had little taste for them.

Often roots were steamed and fed in a mash mixed with bran and a modest quantity of boiled oats or beans. Preparing the mash was entrusted to a day labourer or a stockman who was available to do this during the afternoon while the horses were at work in the fields. 'The horses are exceedingly fond of mash', Henry Stephens noted, 'and when the night arrives for its distribution, show unequivocal symptoms of impatience to receive it.'[47]

By the end of the nineteenth century the quantity of roots fed was far less than had been usual 50 years earlier. Some recommended diets omitted them entirely. Linseed, on the other hand, which had been less-favoured in the mid-nineteenth century was now more firmly recommended, albeit in small amounts.

Winter evenings offered another opportunity for the horsemen to indulge their animals. After the horses had been given their evening meal and grooming, and were put into the stable for the night, the hay racks above each manger were filled. Some farmers laid down rations for the hay, but many did not, and the men could then fill the racks as fully as was possible. John Burke, for one, was unhappy about the opportunities for over-indulgence thus offered to the horses: 'The power and activity of many a team is often diminished through want of supervision by the farmer in the article of diet, and more work might be got out of his horses, besides effecting a great saving of food, did he take as much pains to regulate their allowance of hay or green meat as he does of oats and beans.'[7]

Watering was as important as feeding; indeed it came before food in the order of work. The first task in the horseman's day was to lead the horses to the pond or to the trough for a drink, although in the depths of winter he had to break the ice for them beforehand. In stables which had a cistern in the yard, however, the water might be given in a clean bucket.

At dinner time, and again on returning home in the evening, the horses were led down to the pond before they had their meal. During the day they needed plenty of water, particularly if they had been working hard. The horseman had to ensure, however, that they did not consume too much very cold water immediately after heavy work, when they were still hot, for that might give them colic.

An extra ration.

A drink at the trough. . . .

Or at the pond, after returning from work.

Grooming

What the writers of textbooks recommended, and what the horsemen did, were, of course, two quite different things on many occasions. 'It is not an unusual practice to curry and wisp the horses, and to put the harness on them while engaged with their corn', noted Henry Stephens, 'but this should never be allowed. Let the horses eat their food in peace . . . The harness can be quickly enough put on after the feed is eaten, as well as the curry-comb and brush used, and the mane and tail combed.'[47] This was one stricture that went largely unheeded. More often than not the morning's grooming was done while the horse was busy with its bait.

Grooming had to be done thoroughly, and needed elbow grease. A Yorkshireman recalled a farmer telling one of his horsemen, 'I want to see you wispin' of 'em as though you hated 'em.'[18] A wisp was a simple home-made brush, made from lengths of twisted hay rope bound together. At one time it was virtually the only tool for grooming horses, and it never entirely lost its usefulness. It was recommended, for example, for giving the horses a rub down immediately after their return from work, to pick up the sweat. For most of the vigorous brushing, however, the horseman of the nineteenth and twentieth centuries had bristle brushes. The dandy brush and the body brush were used according to the state of the horse's coat; the dandy brush having longer, stiffer bristles than the other. The curry comb was used to prise out the more intractable pieces of mud stuck in the horse's coat after a day's work, especially in the feather of the legs. It was also useful for cleaning the other two brushes.

Grooming was begun at the head, for approaching the horse from the front always helps to put it at ease. Particular attention had to be paid to the legs and feet, as the accumulation of dirt here would render the horse vulnerable to disease. An unpleasant disease was grease, an inflammation of the skin on the lower part of the legs, especially the hind legs. Brought on by the combination of dirt, damp and cold acting upon the skin, it created a greasy discharge of rather obnoxious smell and caused discomfort to the horse. It was important, therefore, for the legs to be cleaned and dried thoroughly at the end of the day. Horses with a lot of feather were more prone to grease because the hair made it more difficult to keep their legs perfectly clean.

The feet had to be carefully picked out to remove dirt and stones from the frog and the heel. In 1844 John Burke recommended that, to keep the hooves pliable, they should be brushed every second or third day with a mixture of tar and tallow melted together. The soles of the feet, he suggested, should be stopped

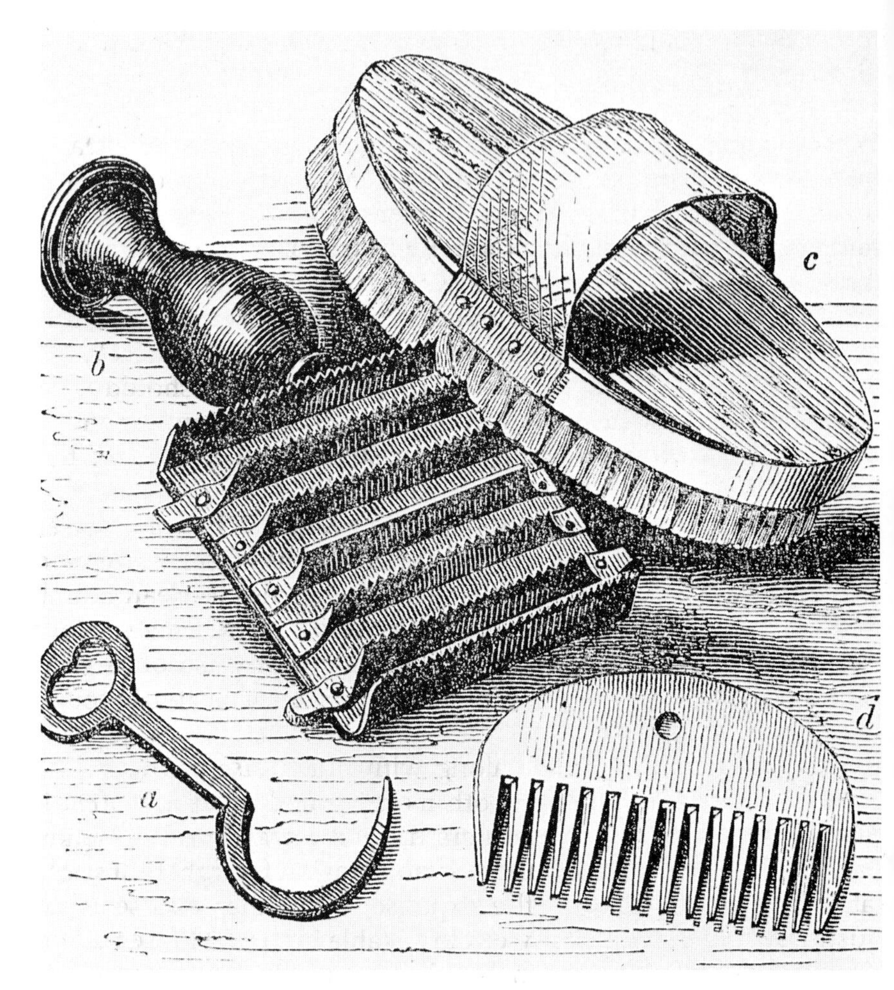

The tools used for grooming; an illustration from Henry Stephens, *The Book of the Farm*. The curry comb is labelled b, the dandy brush c, foot pricker a, and mane comb d.

up each night with soft cow dung, which would prevent the accretion of muck from the stable overnight and make picking out the feet easier in the morning. He had little faith in the horseman's attention to detail, however: 'It is very rarely that a carthorse has his feet cleaned, except when he goes to be shod.'[7]

According to the Young Farmers' Club booklet, at least half an hour was required to groom a pair of horses in the morning. The reality was often different. A retired farmworker, Bill Petch, recalled joining a farm in 1932, and being told by the farmer that no more than two minutes should be spent grooming each horse. The farmer gave a demonstration. The parts covered by the harness were brushed, and 'with luck a few seconds could be spent on the horse's legs.' In practice Bill stretched the grooming out to eight minutes for the pair of horses. When he had to catch and bring in four horses from the field, then feed, groom and harness them, within the hour allowed at the start of the morning, then only a few minutes with the curry comb could be afforded. No harm apparently came to the horses from this treatment; the parts that mattered were brushed thoroughly and kept free of disease.

Bill Petch's experience came when horse-powered farming was on the wane, and only the one man was responsible for the stable of four horses. But such methods were not unusual one hundred years before that time. Henry Stephens complained, 'there are more ways than one of grooming a horse, as may be witnessed by the skimming and careless way in which some ploughmen do it.' The coat of a horse may be rough, he went on, but it should at least be kept clean, though not necessarily sleek. He recommended that the steward or head horseman should keep a proper check on his men's work: 'A slap of the hand upon the horse will soon let you know the existence of the loose dust in the hair.'[47]

Grooming had to be repeated in the evening when the horses arrived home from the fields, and this time special attention needed to be given to the legs and feet in order to clean out all the mud that had worked in during the day. The horses might be taken for a brief walk through the pond to wash off a good part of the mud before using the wisps and brushes for thorough cleaning.

In midwinter this work, along with much else at the stable, would be done in the dark both morning and evening. Often it would be very dark, for farmers did not always provide many lamps. They 'either think there is no occasion for light in a stable at this hour, or grudge the expense; but either excuse is no justification for doing any work in a stable in the dark', remarked Henry Stephens. Those who worked as junior horsemen found, too, that the head horseman had a habit of keeping most of the lights with him.[29; 47]

Keeping the horse's coat sleek and glossy was something that engaged the attention of many horsemen, especially in the larger stables where there might be some rivalry as to who had the finest team. As well as thorough grooming with brush and comb, the horsemen put their faith in a variety of other things that were thought to improve the shine of a horse's coat. There were several herbal recipes, using, among others, tansy, saffron, bryony and gentian, to be added to the horse's feed.[16] There were, however, other more dubious substances. Arsenic was one extreme, and several cases of its use were reported during the late nineteenth century.

Harnessing

The texts usually recommended that there should be a break between feeding and grooming, and harnessing the horse; this was to allow the horse time to digest its meal, and for the horseman to have his own breakfast. Not infrequently, of course, it proved impossible to fit in the break.

To the experienced man 'collaring up' was probably second nature. To the novice it seemed quite a performance. John Stewart Collis recalled his surprise at discovering that a task apparently so simple involved so many operations. He soon found himself in difficulty, trying to put the horse's collar on after the bridle, trying to fit the collar over the horse's head while still the right way up, and putting the harness on the wrong way round.[12] Collis did not have the experience of working through an ordered and hierarchical stable, for these were things which the youngest ploughboys learned early while having little of the 'real' work with horses in the fields.

There was a correct order, of course, which was to put the collar on first and the bridle second. The collar would fit over the horse's head only with great difficulty if the bridle was on, and therefore putting the bridle on second was more comfortable and calming for the horse. It was also argued that if the horse was being uncooperative in the morning, there would be little damage if it ran off with only its collar on.

The collar had to be slipped upside down over the horse's head, since the collar had to be wider at the throat than at the neck, whereas the horse's head was the opposite, wider across the eyes than the nose. The collar, therefore, had to go on the wrong way up and be turned round on the horse's neck, followed by smoothing the mane down comfortably.

A fair amount of cleaning and grooming looks to be the prospect at the end of the day for these horses in charge of the liquid manure distributor.

Putting the collar on brought another surprise to the beginner, for it was remarkably heavy. Collars for heavy horses commonly weighed 12 to 14 lb in the southern counties of England. Further north they tended to become heavier, until in Scotland weights of 20–30 lb were not unusual. These heavyweights were mainly the Scottish peaked collars on which the sides had extra stiffening with whalebone, and the cape on the top of the collar was brought to a sharply pointed peak high above the neck. English horsemen rarely could see the sense in the high-peaked collar, which did tend to be more decorative than useful, and they thought the weight rather excessive, to say the least.

The collar was arguably the most important single item of harness for the draught horse. Its purpose was to attach the horse to its draught, making use of the strength of the horse's shoulders, but at the same time enabling the horse to take its load in comfort. The collar therefore was heavily padded to avoid soreness and to protect the horse from undue pressure, especially on its windpipe.

There were three main parts to the neck collar. The forewale was a tube of leather filled with straw, packed in hard to make the tube stiff. Attached to this was the body which acted as the padding of the collar. The padding was of straw, rye straw by preference, and was covered with a thick woollen cloth that almost always had a wide check pattern, and thus was known as collar check. The padding was thicker at the points of greatest wear, especially at the points of draught where the chains were attached. The whole of the body was covered with broad leather side pieces—the afterwale—to protect it from the weather and wear from the chains.

A Scottish haymaking scene with a horse wearing a peaked collar. The Scottish pattern of cart saddle, with the chine built up at the front more than most English ones were, is also evident here. The illustration is from a postcard issued in the early twentieth century by one of the Scottish publishers prominent at this time.

A collar with brass hames. It was used on the Duke of Wellington's estate.

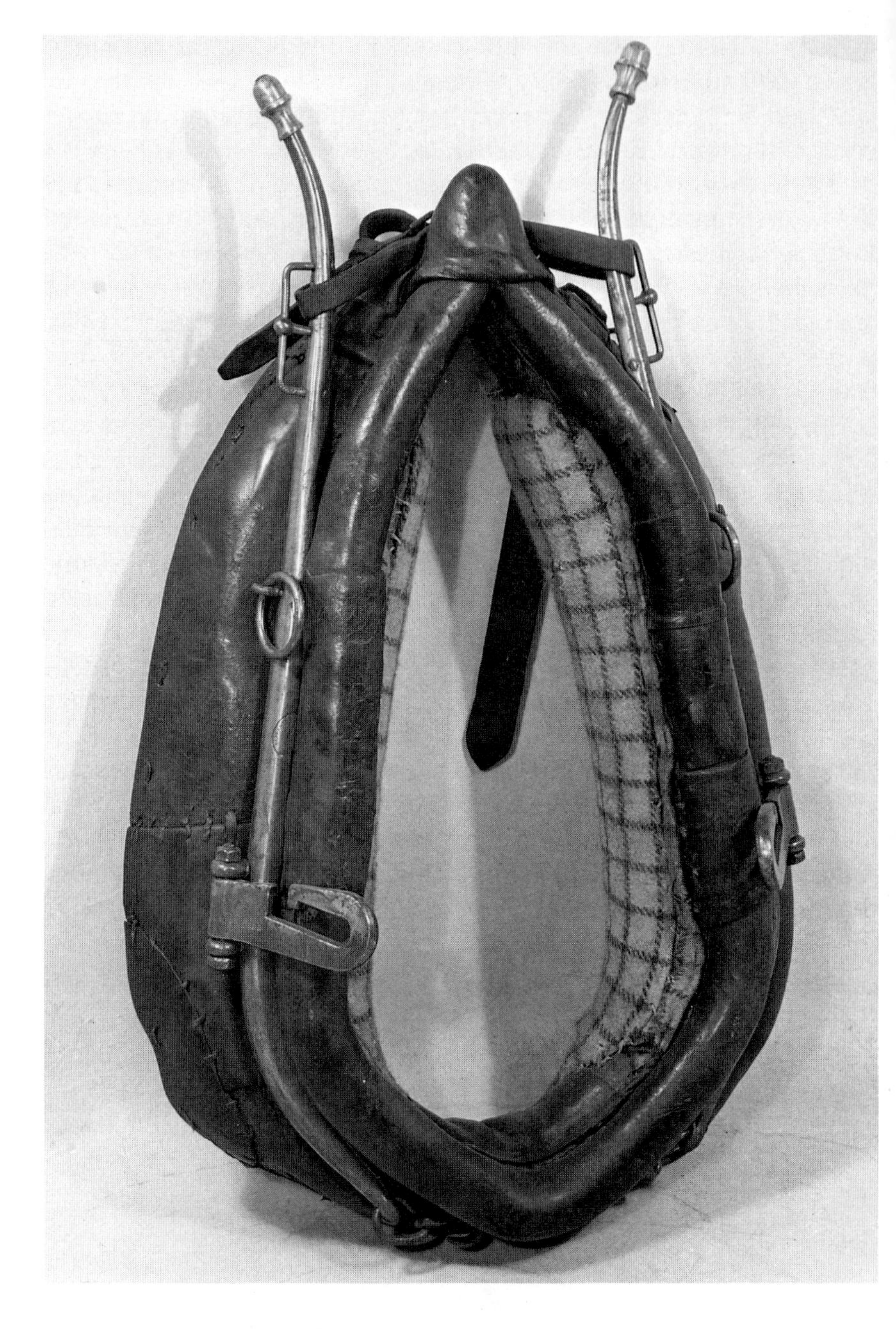

A cape or housen was often attached to the top of the collar. This was a stiffened piece of leather to 'prevent rain from getting between the cushion and the shoulder, there to heat, irritate and even to blister the skin', as the *Farmers' Dictionary* of the 1840s described it.[19] The best housens were the simplest: rectangles of leather which could be made to lie flat above the horse's shoulder during wet weather or stood upright to allow air to circulate when it was dry. There were many other designs, especially during the early and mid-nineteenth century, which involved large sheets of leather, often semi-circular, or nearly so, and which tended to stand proud, like a sail. They were an English peculiarity, especially in the Midlands, and smaller versions remained common, but more as ornament than for any practical purpose. Henry Stephens thought this type of cape 'answers no purpose of protection from rain, but rather to catch the wind, and thereby obstruct the progress of the horse. Such a cape is frequently ornamented with flaring red worsted fringes round the edge, or with large tassels from the corner and middle, and even with bells.'[47]

Next to be put on were the hames, which formed a frame around the collar to which the ropes or chains of the harness were attached. Hames were originally made of wood, ash or beech for preference, and were painted for protection. Wooden hames remained in use in some areas until the end of horse power on the farm, but during the nineteenth century, iron ones became more common, and were the usual form by the century's end. Further developments introduced hames of steel and brass, and while these were often regarded as being for show purposes only, it was certainly not unknown for farmers to use them in everyday work. Hooks over which the chains could be attached were fitted to the hames. Most hames had hooks suitable for all normal purposes. Some, though, were designed for specific jobs. Shaft hames had fixed draught chains for the horse working in shafts. Body hames had chains and hooks fitted in position for the horse harnessed in the middle of a line of three or four. These specialised hames were, nonetheless, sometimes used for general purposes.

'The haims were never removed from the collar,' declared Henry Stephens. While many horsemen accepted this conveni-ence, the practice of taking the hames off each night was at least as common. In the Midlands, for example, horsemen argued

Four horses ploughing in line, each with its housen raised. Working with four in line usually required a driver, as well as the ploughman. The driver had to ensure that the horses walked the correct line, along the bottom of the furrow. These horses seem to be showing no sign of straying, hence the relaxed distance at which the driver walks.

that it made the collar uncomfortable to leave the hames on and therefore removed them each night. The hames were fitted again each morning, put around the collar and made fast with a short chain at the bottom.

The next task, according to the texts, was to put the bridle on. There were local patterns of bridle, but the major distinction was between the closed bridle, that is with blinkers, and the open, without blinkers. There was heated debate during the nineteenth century as to the merits or otherwise of restricting the view of the horse, and impassioned arguments against the use of blinkers. In practice much depended, as Henry Stephens argued, on what the horse was used to. The horse accustomed to blinkers was very likely to take fright at the slightest distraction. 'But horses broke in', he continued, 'without the bridle, are less likely to be scared by any occurrence in the road, than those accustomed to it, as they see every object near them distinctly; and the want of it keeps the head cooler in summer, and saves the eyes injury by its pressure.'[47] The effect of any such argument on the world of agriculture, however, was limited, for the majority of horses on the farm worked with blinkers.

The uncommon choice: the bridle without blinkers, worn by one of the horses on the Webster Cory estate at Notgrove, Gloucestershire in the 1930s. This farming enterprise specialised in dairying and poultry, and employed light horses for all the carting around the estate, while tractors were used for the heavy field work, such as haymaking.

While some horsemen preferred to put on the straps, saddle and breechings before the bridle, others left these until last. What harness was worn depended on the work which the horse was to do. Cart harness was worn by horses working in shafts,

whether in cart or wagon, or in field implements such as the roller, horse hoe and fertiliser distributor. Plough harness, or chain harness, was worn when drawing any implement without shafts, which, in addition to ploughs, included harrows, binders, most types of seed drill and carts which had a central pole in place of the shafts. Trace harness was for the leading, or trace, horse when a wagon or implement was drawn by two horses in line.

For cart harness a cart saddle, or cart pad, was worn. Its function was to take some of the weight of the vehicle, making it more evenly balanced and less of a strain for the horse. The saddle, therefore, had to be well padded in order to sit comfortably on the horse's back and ribs, while not putting pressure on the spine and withers. There were various patterns for the cart saddle, but the essential elements were the same. The tree, usually of beech or elm, was what really took the load, and was attached to the pad, which was stuffed with straw and lined with collar check. The whole saddle might then be covered with a leather housing.

The saddle was held on by a girth band. Straps along the back of the horse—the meeter strap attached to the collar and harness, and the crupper strap along the horse's rump—kept the saddle from slipping backwards or forwards. A breech band (or breechings) was fitted around the horse's hind end, attached to the crupper by hip and loin straps, acting as a buffer when

Wagons in Yorkshire were often built with a central pole rather than shafts, to which the horses could be hitched using plough chains. This posed photograph of the 1900s shows the wagoner riding postillion, and with a riding saddle, a comfort not all preferred to have.

Cart harness worn by a horse in Norfolk. The simple form of cart pad, with no leather housing, was common in East Anglia and southeastern England. This photograph is another illustration of the muddiness of much of the horse's work.

bringing the wagon to a halt or pushing it backwards.

Trace harness was simpler, since neither the breech band nor the cart pad was necessary. Instead of the pad a broad back band was worn, which had hooks to take up the chains of the harness. A crupper strap again held the back band in place, and also supported the hip strap which held the chains at the hind end of the horse. Trace crupper straps usually ended in a dock around the horse's tail. This dock was made of soft leather, and filled with linseed. Heat from the horse's body would warm the linseed causing the oil to run inside the dock, making it more supple and comfortable. Immediately behind the horse a spreader bar was fitted between the two harness chains to keep them apart and away from entanglement with the horse.

Plough harness was simpler still. A plough band, or back band, was all that was needed to take up the chains between swingle-trees and collar. In the Midlands plough harness was commonly known as G.O. gears or G.O. tack. This apparently derived from the ploughman's call 'Gee oh' to turn the horse to the right as it reached the end of the furrow. But in Hertfordshire trace harness was called G.O. tackle.

It was always recommended practice to have a separate tack room for the harness. Many farms did not, and the harness was kept in the stables, hung up on a wall, or even on the posts by the stalls. Each horse had its own collar, bridle, trees and halter. Ideally, the collar should have been made for the horse and shaped and moulded to fit around its neck with comfort, but as with many things, this was not always the case. There should then have been sets of the different types of harness, a set of cart harness and of plough harness for each horse, and sufficient sets of trace harness—one set to every three or four horses—as well as many pairs of reins, plough strings and a number of riding saddles. Leather harness was expensive, however, and the farm in Suffolk on which Hugh Barrett trained in the 1930s was not unusual. The harness had been bought piecemeal, most of it second-hand, and did not perfectly match its fellows, and there was no guarantee of finding the hames or other harness needed for particular work. Things could have been worse: a neighbour apparently had no breeching straps at all for his wagon harness.[5]

As far as the working horse of the farm was concerned, the practice of adding decoration to the harness grew up during the eighteenth century. At first this took the form of using decorative thonging and tooling on the leather of the harness. It became common to have a simple pattern and the date tooled into the harness, especially on the housen. Another form of decoration sometimes used in the first half of the nineteenth century was the use of brass nails on cart saddles to secure the leather housing to the tree. More colourful decoration was found in the woollen fringes and tassles that were attached to the housen. Bright colours, of red, blue and yellow, seem to have been favoured. These decorations remained popular into the mid-nineteenth century, and have survived to the present day in some show harness.

Brass ornaments first appeared at the end of the eighteenth century. The early ones were in the form of small ovals for the blinkers or to go on a face piece. During the course of the nineteenth century, brasses gradually grew in popularity as a form of harness decoration. Brass foundries, many of which

Trace harness and cart harness, from a saddler's catalogue of the late nineteenth century.

Plough harness, illustrated in J. C. Morton, *Cyclopedia of Agriculture* (1856).

There were many attempts at making improvements to harness, although few gained much popularity, however good they might have been. Owen's patent harness was advertised in the *Implement and Machinery Review* in 1905.

were located in the Midlands around Walsall, produced brasses very cheaply and offered a wide range of designs, and this undoubtedly helped this form of decoration to spread. The fashion reached its peak in the late nineteenth century, falling away during the 1920s, as horse power began to wane.

On the farm it was the horseman himself who usually collected his own brasses, rather than their being part of the farm's turn-out. In ordinary daily work brasses were not usually worn, although some horsemen liked to have a face piece or a breastplate with a few brasses on. It was for the special occasions, especially ploughing matches, that the farm horses were really decorated, and then all of the horseman's collection came out. There were not places enough on the ordinary plough harness to attach the brasses, so that extra straps were added for the decoration. These varied considerably from region to region, and especially between England and Scotland. In England it was common to have one running down the side of the neck to the bearing rein, and a breast plate attached to the collar and taken between the fore legs to be joined at the girth strap. This breast plate was usually large enough to accommodate four brasses. There were also decorative side straps, most commonly one before and one behind the saddle or plough band. Those who went in for decoration in a grand way would find more places along the horse's flanks for side straps and ornamental chains. Decorative terrets might also be worn, the most usual being one on the crown (the fly terret) and one on the cart saddle.

Brasses were likely to be worn when taking the horses to town with a wagonload of corn, since both horseman and farmer liked to see a good turn-out before public gaze. Village festivities, such as harvest home, harvest festivals, and Sunday school outings, were other occasions when the full decorations were brought out.

Shaft Work

Harnessing the horse to a cart or wagon went by a number of different names: shutting in, shooting in, hitching in and others. It was another task that could surprise the inexperienced. It could not be taken for granted that the horse was going to cooperate with new hands and walk back into the shafts. Once the horse was there, the shafts had to be held up with one hand while throwing the chains across with the other, followed by hitching the chains to the correct hooks. It was important not to have the chains so slack that the shafts were held low on the horse's side because then the load would be below the horse's centre of gravity and thus harder to pull.

The horse that was to work in shafts was known across much

Driving the carts out along an avenue of chestnut trees. This was a common driving position, again on the nearside of the horse.

of southern England as a thiller, or thil horse. There were many other local names. In Kent and Sussex, for example, it was quiller, or coiler. The trace horse, leading a team in line, was widely known as the fore horse, which in parts of eastern England had been corrupted into foorest, or even forrest. In the line of three, the middle horse was, in southern counties again, called the body.

Having got the horse hitched in, the next challenge was to negotiate the tracks and roads successfully, for there were all

Village agricultural shows, once a common feature of country life, were occasions when horses were displayed in their greatest finery. These, showing almost every kind of decoration, were at the show in Little Weighton, East Riding of Yorkshire, in 1912.

The team dressed up for the ploughing match, with terrets on the head, a breastplate full of brasses, ornaments on the housen and blinkers, and polished and decorated chains and straps.

sorts of hazards, some on the road and some arising from the temperament of the horse, which are largely no longer a problem with diesel power. Gateways always needed care. The booklet *Farm Horses* emphasised that they should be approached square on. Coming at an angle was likely to be unsuccessful, and the many horsemen's recollections of mishaps at gateways affirm that. Having hit several gateposts, the young Bill Petch, in his first job, was told by the carter to leave it to the old mare, who was far more experienced. It worked. Steering with a trace horse, and with a wagon, longer and heavier than a cart, needed extra care. Sharp dips and rises, and narrow tracks could also be hazardous.

Yoking to the Plough

When ploughing with horses in line it was usual to fit a spreader bar between the chains behind the back horse. From the middle of the bar a single chain linked with the beam of the plough, and this was enough for the draught power of the horses to be transmitted to the implement. When horses were worked abreast, a means had to be found whereby the power of the individual horses spread out in front of the plough could be combined and concentrated at one point, the beam of the plough. This was achieved by use of the form of drawbar known variously as swingletrees, whippletrees, bodkins and several other local names.

Gently through the gate. A laden hay wain in Somerset, about 1900.

Swingletrees traditionally had been made of wood, usually oak, but Henry Stephens in 1849 noted that there had been some successful trials of malleable iron. The iron trees had one big disadvantage; they cost 16*s* compared with 12*s* for the wooden ones. By the 1930s, though, iron swingletrees were quite common.

Working with two horses abreast required a simple set of three swingletrees, one behind each horse, these being linked to the ends of a central tree that connected with the plough (Figure 1).

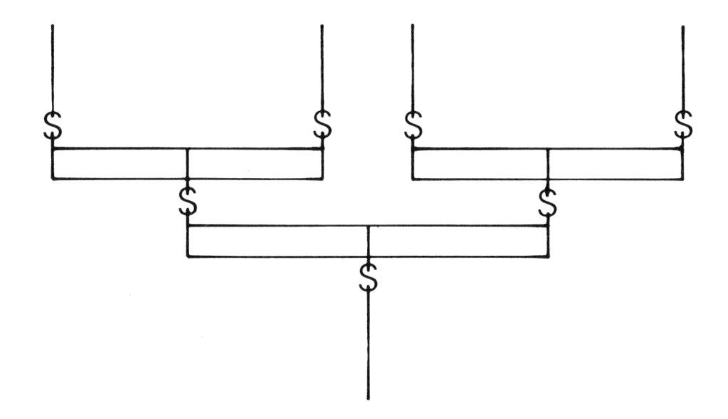

Figure 1 Yoking for two horses abreast

The horses were usually coupled together with a rope linking the bits or the hames to ensure that the horses pulled together rather than one trying to get ahead of its partner. This became the most common yoking for ploughing with the single furrow plough during the nineteenth and twentieth centuries. The left-hand horse walked along the unploughed ground, and was known as the land horse. The right-hand one was the furrow horse, for it walked along the bottom of the last furrow ploughed. In this way the horses' feet hardly touched the newly ploughed ground.

When three or four horses were yoked together arranging the drawbars could be more complicated. There were many patent arrangements of swingletrees, but the most common way to couple three horses abreast was that shown in Figure 2. The off-centred coupling to the plough, placed one-third of the way along the main bar, combined with the yoking together of two of the horses through an additional swingletree, to produce a simple two-to-one ratio that equalised the draught of the three horses. This yoking could be arranged either left-handed, as shown in the diagram, or right-handed. When this harness was used with a plough, it was usual to have the one horse walking in the furrow, the two on the left on the land.

An improved arrangement for three horses was the use of compensation levers to link three simple swingletrees to the main drawbar. These iron levers equalised the pull of each horse on the drawbar, which now could be attached at its centre to the

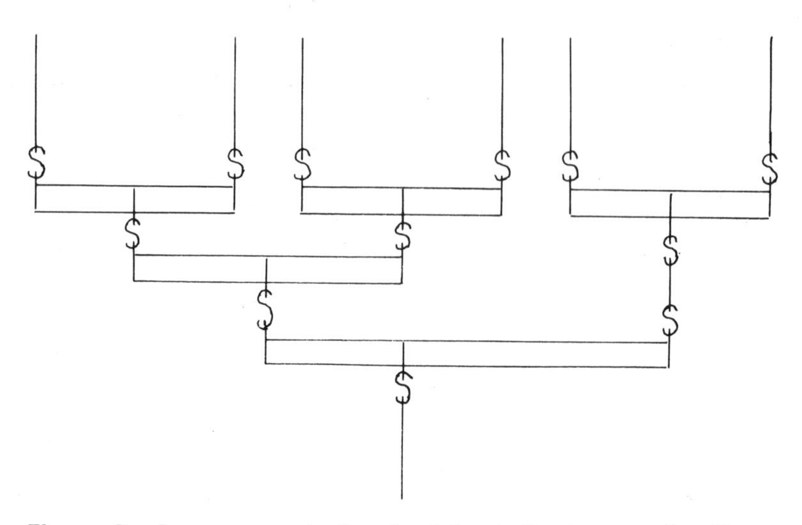

Figure 2 Arrangement of swingletrees for harnessing three horses abreast

plough (Figure 3). The levers could be adjusted by moving the bolts at the fulcrum so that the load on individual horses could be altered. If a horse in a team of three was noticeably stronger than the others it could thereby be given a greater share of the draught. Alternatively, a weaker horse, perhaps a young horse still new to harness work, could be given a lighter load. Compensation levers appeared to have every advantage, except that they were considerably more expensive. And as many of the compensatory adjustments could be made, if really desired, to the more simple three-horse yoke, the compensation levers were used less commonly than they might have been.

There were other ways of yoking three horses. One was known as unicorn fashion, where the middle horse walked ahead of the other two. A simple set of swingletrees was all that was needed for this arrangement, with a long chain to attach the middle horse

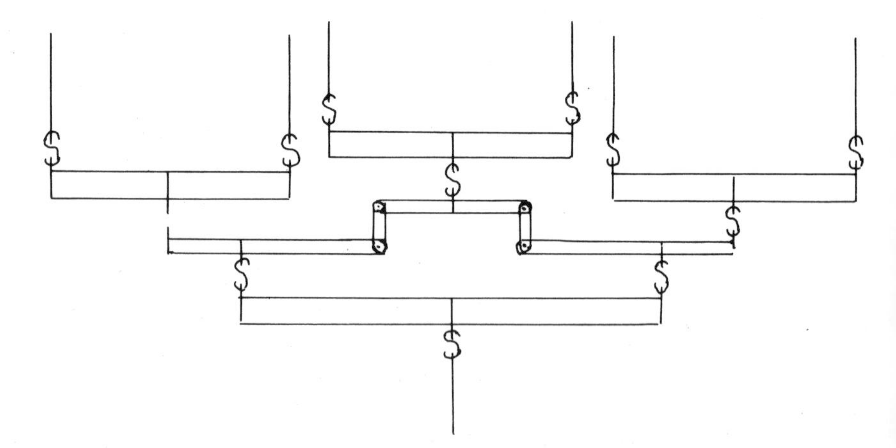

Figure 3 Three-horse yoking with compensation levers

to the drawbar (Figure 4). It was a neat arrangement which kept all the power pulling forward, whereas with three horses abreast on a plough some of the energy was drawn off in sideways pull. The difficulty with unicorn yoking was that there was no way of equalising the draught of the middle horse.

Three horses worked bodkin fashion had the third horse in front, but in line with the right horse of the pair abreast, linked by simple trace harness. For ploughing this meant that two horses would walk in the open furrow, and this was practised by some farmers in Yorkshire, Lincolnshire and Norfolk. As a means of yoking to other implements, such as binders, bodkin fashion was a little more widespread.

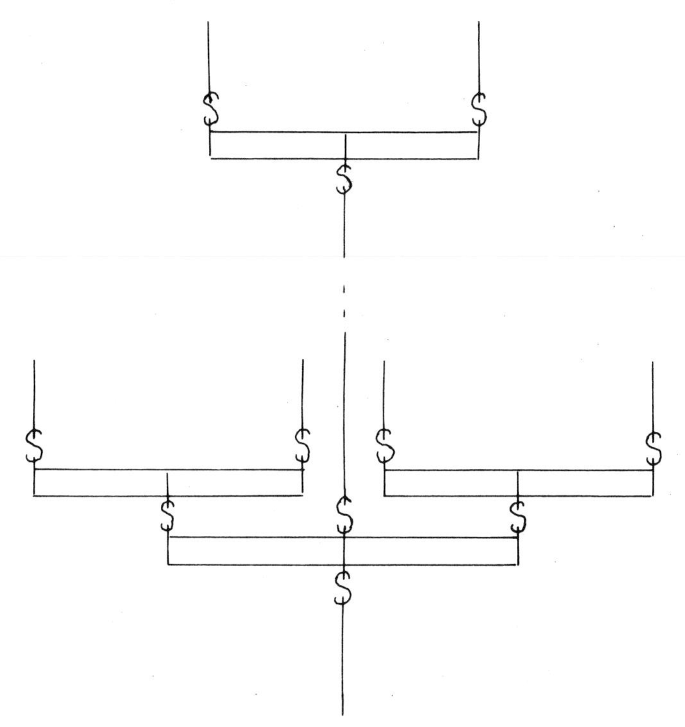

Figure 4 'Unicorn fashion' of harnessing three horses

There were various ways of yoking four horses together. Four might be worked abreast, to pull the heaviest of harrows, for example. They could then be harnessed using an extension of the swingletrees for three horses. Four horses could also be worked in two pairs. For that, the simplest method of yoking was an adaption of the unicorn fashion for three horses. The leading pair was simply linked to the main drawbar by the long soam chain. This suffered again from the problem that it was not possible to ensure that the two pairs worked equally hard. The arrangement illustrated in Figure 5 shows one solution adopted. By running the chain round the pulley behind the drawbar, the two pairs were made to pull at opposite ends of it, which had the effect of

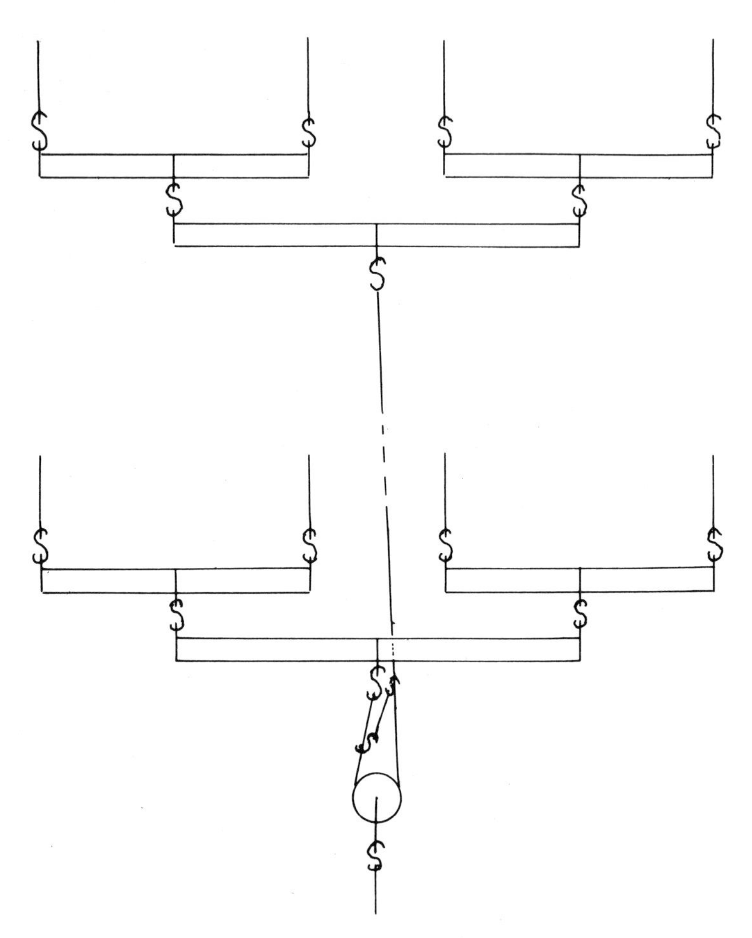

Figure 5 Yoking for four horses with equalising pulley

equalising the draught. Henry Stephens claimed that the simple linking of the soam chain was 'now seldom employed', being superseded by this or other means of equalising the work. But the simple yoking retained its attractions for many, even though less efficient.

The horses were driven by the reins, lightly pulled on the bit, as when ridden. In the Fens of Lincolnshire and Cambridgeshire it was the custom to drive a team of horses with only a single rein instead of the conventional pair. Instead, they had a looped rope, known as a whip-line, attached to the left handle of the plough. From it a cord ran along the horse's left flank to the bridle. A light tug or a shaking of the cord steered the horse to left or right. The left-hand horse of a pair had the line, the middle one of a team of three. The coupling chains or bars between horses ensured that the line horse pulled or pushed the others around.

People from other districts were often mystified as to how the horses could be kept properly in control with only the one rein. Perhaps it owed something to the horseman's renowned habit of talking to his animals. That, however, would not do for Henry

(*Above*)
Three horses to the plough, Berkshire, in the 1930s. No back band is being worn, quite a common practice.

A conventional pair of reins being used to drive a team of three with a harrow on a farm in Norfolk in the 1930s.

Three horses harnessed unicorn fashion to a binder, and with a lad riding postillion on the leading horse. The photograph was taken in Yorkshire in the 1900s, and the long-strawed wheat then grown is readily apparent here.

Stephens. 'It is not practicable to make horses at the plough go through the proper motions with a single rein', he wrote, 'nor is the single rein at all commendable, inasmuch as ploughmen accustomed to it, fall into the practice of incessantly bawling to their horses, which at length become regardless of the noise, and make the turns at their own leisure.'[47]

Talking to the horses was as important as using the reins to drive the horses, and a large number of dialect words were used. There were some common features. 'Whoa', with variations here and there, was almost universal to stop a horse. To start it, clicking the tongue, or calling 'gee up' was common, and to get it to go backwards most men called 'back', 'hup back' or a variation on that. When it came to turning to right or left, dialect came into its own. The Lincolnshire ploughmen with single reins were likely to bawl 'gee' to go right, and 'arve' to turn left. In Suffolk 'woordie' was right, and 'cappa-ree' was left, and there were numerous variants. Elsewhere, turning right might be 'hup', 'gee again', 'gee-hoe'; and to go left, there was 'hie' (especially in northern England and southern Scotland), 'heck' and 'cum-ere'.

(*Below*)
This shows one way of harnessing four horses to a binder. A piece of sacking is all that is afforded the boy as a saddle.

(*Above*)
The four horses on this drill are yoked as two trace pairs.

Three horses making steady progress in a wide, open field, about 1920.

Of all the tasks in the horseman's year, ploughing was the most intensive in terms of both his labour and that of the horses. Ploughing with horses could never be performed with speed. The most that could be ploughed in a day was one and a half to two acres, which was reckoned to be common across much of Norfolk at the end of the eighteenth century, an achievement that Arthur Young put down to the ploughmen, 'who have been accustomed to keep their horses and themselves to a quick step, instead of the slow one common in almost every other district.'[57] In neighbouring Suffolk no more than three-quarters of an acre was common on the heavier lands. The rate of work depended on several things: speed at which man and horse walked; lightness of the soil; quality of the plough and its resistance to draught; and size of the field, for in a very small field about half the time could be spent turning at the headlands. While the rate of ploughing varied from district to district, an acre a day was the standard reckoning for most practical purposes. That represented a lot of walking: more than 12 miles, according to calculations of the mid-nineteenth century recorded by Henry Stephens.

Ploughing thus represented a large part of the labour of horse-powered farming. In the arable eastern and southern parts of the country, where four-fifths of the land could be in tillage, for the farm of 100 acres, about 80 days were in effect taken up in ploughing, and that was excluding any second or cross ploughings. Of course, the period was more compressed if two or more plough teams were at work, but the measure of the task is clear. Ploughing accounted almost entirely for the peak of work in the autumn, and contributed appreciably to the second peak in the spring, when some land was being ploughed for late-sown crops, such as roots.

It is hardly surprising that ploughing should have played a dominant part in the horseman's working life. Indeed, he was employed at least as much for his skills as a ploughman as for his abilities at handling horses, although both usually went hand in hand, as the most skilled ploughmen needed the lightest touch on the reins to keep the horses in a perfectly straight line. A favourite trick of the expert ploughmen was to demonstrate that they could set the horses going, walk away and take up the reins again near the end of the furrow. A combination of steady, experienced horses and a plough with its wheels set at just the right angle was the secret, but these exhibitions provided the stuff of countless stories of the ploughmen's exploits.

Senior in this, as in all things, the head horseman took the lead in ploughing a new field. The first task, except when a turnwrest or one-way plough was being used, was to mark out the lands or stetches into which the field was being ploughed. The stetches were sets of furrows turned into a ridge at the centre, and falling away to a deep furrow, or 'water cut' at the edge which joined neighbouring stetches and also acted as a drainage course. There were set widths to these stetches: narrow (two yards) for heavy land that needed more drainage, and wider, three yards, for lighter soils. The spread of stetches across the field had usually been long-established, but the ploughman still had to ensure that the positions were precisely marked out again. If this was not done the required number of furrows would not fit properly, which would not only throw out the farmer's calculations for subsequent work, but was a source of injured pride to the ploughmen. Hazel sticks therefore were positioned to mark exactly the line of the ridge furrow of each stetch, and to act as sight lines for opening out the first furrow. To the most skilled ploughman, who as often as not was the head horseman, fell the task of drawing the first furrow, for he was the person most able to draw a perfectly straight furrow precisely in line with the hazel marker, and thus set the stage for a perfectly ploughed field.

Skill in ploughing was of great importance, also something which came with experience and, for beginners, encouragement. Young men who were learning to plough soon found that they did not have enough hands to hold on to the handles of the

The stetches can be seen neatly marked in this large field on chalk downland. Three teams at work represent some considerable investment into the effort of ploughing.

plough and at the same time control the reins of the horses to keep them from wandering out of line. The good horsemen took the young lads under their wing. Rarely would they be allowed to handle the plough until they were 17 or 18; younger than that they were not strong enough to control the plough. In learning, one of the skilled men would lead the way, while the learner, with one of the more placid pairs of horses, followed.

An experienced man might follow the learner to cover his mistakes, for neatness in ploughing was more than a valuable practical skill, it was a matter of pride, a measure of the standards of a horseman's work and of the farm on which he worked. Fields next to the road, in particular, had to be perfect, for they were on public view. This was especially so on Sunday afternoons when ploughmen were likely to spend their day off walking round the lanes inspecting the work of their neighbours and rivals.

The tradition of measuring a man's skill by his ability to plough a perfectly straight furrow developed during the eighteenth century. It was most fully established in Suffolk at that time, universally recognised as the English county where the highest standards of tillage were to be found. Arthur Young remarked on the Suffolk men at some length: 'The ploughmen are remarkable for straight furrows; and also for drawing them by the eye to any object, usually a stick whitened by peeling . . . and a favourite amusement is ploughing such furrows, as candidates for a hat, or pair of breeches, given by alehouse-keepers, or subscribed among themselves, as a prize for the straightest furrow.'[59]

The larger, more formal ploughing matches had already come into being by the end of the eighteenth century, and these became of increasing importance during the nineteenth century as occasions when horsemen could demonstrate their expertise and establish a reputation throughout the district—a reputation that could be cashed in when the man considered moving to another farm. Turn-out was everything on these occasions, and for days before the match the horseman would be hard at work preparing the plough, cleaning and oiling the harness and polishing the brasses. On the day itself, the horse was given its most thorough grooming of the year.

While ploughing took up the greater part of the horseman's time during the autumn, the season had other tasks to be performed. Potatoes were being lifted, and the horseman had to lead with the potato-raising plough or the potato spinner to open up the ridges while the gang of pickers followed behind gathering the crop by hand. The other root crops, turnips, swedes, and, later, sugar beet, were lifted at this time as well, and all of these had to be carted to the clamps where they were stored. Meanwhile the seed drills were out getting the winter wheat and beans sown.

(*Facing page*) 'Nothing can really compare with the strenuous horse-work', John Stewart Collis observed of ploughing. And a fine team such as this, seen at a ploughing match in Surrey in 1935, demonstrates why he might have thought that.

(Above and facing page)
The busy scene of the ploughing match. Almost all ploughmen had a matching pair of horses. They knew that improved their chances in the judging for turnout, whatever the theory of the rules might be.

Ploughing had to be finished by Christmas. That was the tradition, and one that had a practical point, for the ground was likely to be harder by January. After Christmas things were quieter for a few weeks. The horses could have some rest while the horsemen might assist with threshing and other general work about the farm, although the head horseman, as befitted his status, was often exempt from such labours. There were likely to be several winter days when snow, deep frost or heavy rain prevented arable work and the horses had to be confined to the stables. A strong school of thought held that it was bad for the horses to be completely lazy, and therefore it was not uncommon for farmers to insist on the horses being taken out for a little exercise even in bad weather.

Nonetheless there was work for the horses even during the winter months. Not for nothing was the horseman so often known as the carter or the wagoner, for the amount of haulage work in farming was considerable. It went on all year round, and the time worked by the horsemen and their teams on carting throughout the year probably fell not far short of that given over to ploughing. There was muck to be taken out to spread on the stubbles, crops of potatoes and turnips to be carted to the clamps, feed to be taken to the fields where sheep were folded, and bags of seed corn to be taken out to replenish the drills. During haytime and harvest, of course, the wagons were fully occupied carting to the stacks. Finally there was road work, taking produce off the farm and bringing supplies back.

During the winter the journeys to the millers, the corn and cake merchants in the towns and the railway station were a regular part of the horseman's work. As the stacks of corn were threshed, a little procession of wagons set out to take the corn away. There was also a regular trip, often once a week, to fetch supplies of cattle cake. Coal also had to be fetched from the railway station,

often in some quantity, as farmers usually catered not only for themselves but for all their labourers as well.

Farmers who were poorly placed in relation to towns or railway stations had to send their wagoners off on long journeys. Distances of 14 and 16 miles were perhaps unusual, but one farmer in Scotland was that far from his nearest railway station and from the pit where he bought his coal. For the return trips of up to 32 miles, his teams would be on the road for between 12 and 14 hours.[36] For the majority of farmers, journeys were shorter, but farms between five and ten miles distant from town or railway station were not uncommon. Regardless of how short the journey, haulage could be heavy work on difficult roads. Since three tons or more of coal or corn could be carried, it was often necessary to assign three or four horses to each wagon.

Trips to the town and the station were often occasions for a bit of a show. An old wagoner from Yorkshire recalled the 'grand sight' of five heavily laden wagons of corn being drawn away, each by four horses with their tails plaited and polished brasses on the harness.[13] There were farmers, and head horsemen, who would make a point of harnessing four horses to the wagon even

A pair of Clydesdales draw a potato spinner on the island of Arran.

(*Below*)
A threshing scene before the First World War, with two horses ready to take a wagon loaded with grain away. Both horses are smartly turned out to have their photograph taken, and are wearing brass face pieces. The stacker here was linked to the threshing machine, and driven by the steam engine, so no horse was needed to work the horse gear.

(*Above*)
A crop of turnips being unloaded into the clamps on a farm in Lanarkshire in the autumn of 1931. Single-horse carts were almost invariably used for this work.

Taking the hay out to the fields for winter feed was work for two horses with the wagon.

A load of hurdles being brought out to the sheep folds on a downland farm.

when two would have been adequate since the horses were on show, not just to the people of the town, but, even more importantly, to the horsemen of the neighbouring farms who would all be making similar journeys. As they gathered in the yard at the railway station the men inspected each other's teams, so they all wanted to have their own horses at their best.

These journeys had their own special excitements and hazards, for the sights, sounds and scents of the town always appeared novel to the horses, and the men always had to guard against distractions leading the horses aside or of the horses taking fright and stopping dead in the street. Horses had many different foibles: they might take fright at double-decker buses, or refuse to go under railway bridges unless completely blindfolded.

Springtime brought another build-up of work, following the trough of January and February. There was more ploughing to be done, as well as cultivating, harrowing and rolling, in preparing a seed bed for the spring corn and clover. Ten acres a day was reckoned to be the usual amount of harrowing that could be accomplished. Although lighter work than ploughing, some of the large chain harrows required a lot of energy to pull them, and two or three horses were often used. It was also somewhat dull work—even the horse was bored, John Stewart Collis believed.[12]

Drilling was also heavy work. Many of the big corn drills needed three, even four horses to pull them, and two or three men in attendance. It was precision work to set good straight

A Cotswold wagon with a load of brushwood faggots, in about 1900. Double-shafted wagons such as this were uncommon, and it was thus unusual to harness horses abreast to a wagon, except in Yorkshire, where the central pole was adopted.

drills, and there was often a rush to get all the corn drilled during late March and early April, so the horseman had to push his horses to their limits.[13]

Late spring and early summer could be relatively unhurried. There were the cultivations to be done in preparation for the sowing of root crops. There were the after cultivations—rolling the early growth of cereals, chain harrowing of clovers and top dressings of fertiliser. Horse hoeing was also important. It was common to have one or two horses especially reserved for this job—the steadiest horses that could walk between the rows and not hit a single plant.

Haymaking was the climax of activities in the early summer. Until the second half of the nineteenth century work for the

(text continued on page 98)

The more usual way of increasing the horse power attached to a wagon was by harnessing in line. Four horses pull a load of timber from a muddy field into the lane in Herefordshire in about 1935.

(Below)
Distractions for the horses came in many forms. The hunt could be an unwelcome sight for the ploughman, his horses were likely to want to join the excitement. Even the most docile could become difficult to hold back, and occasionally they did break their traces.

A heavy job for the horses: harrowing a steep hillside in Denbighshire during the Second World War.

Four horses with a broad zig-zag harrow on the vast fields of the Berkshire Downs. The horses are harnessed by individual drawbars, rather than multiple swingletrees for working abreast.

An attempt to simulate the tractor: five horses and a gang of three Cambridge rollers. Bales of straw add a little extra downward pressure, which would normally be exerted by the driver in the seat. The main purpose of this exercise, in the 1930s, was presumably to save labour, as one man is employed where there would be three in more conventional harnessing.

Drilling the broad acres of southern England with drills worked in multiple—the only way to get the crop sown on schedule on such large farms. Although the horses appear in sole charge here, inspection with a magnifying glass reveals the men's hats.

Horses did not always walk in straight lines across the field. To compensate, many seed drills were fitted with a forecarriage which could be steered independently. The horses were harnessed by simple plough chains on to the forecarriage, as this type of drill usually did not have shafts or a central pole. The steersman, often a boy, looked after the course of the drill, while the horses were driven, in this scene, by the man at the back, who also watched the operation of the drill.

The more usual way of working rollers with horses, one roller drawn by a pair. The ground is being prepared for grass or clover to be sown from the broadcasting barrow (or 'shandy' barrow) during early summer.

(*Right*)
A pair of Percherons with the horse hoe in a field of sugar beet on the Fens, near Peterborough.

(*Below right*)
Lighter power, by comparison: a pony with a horse hoe in the 1920s.

(*Facing page, top*)
Taking the mowing machine out to the hay field, early in the present century. The photograph was taken in Northumberland where Scottish influence was strong—horses of Clydesdale type, and the high peaked collar. All, including the horse on the van, pay full attention to the camera.

(*Facing page, bottom*)
A fine turnout for mowing the hay. Rutland, early 1950s.

A busy haymaking scene in the 1930s. Two horse-drawn hay sweeps are gathering in the crop, as well as a horse rake, while another horse turns the gear to drive the stacker.

The horse plods round and round to keep the gear turning and the elevator feeding the sheaves on to the stack.

A harvest scene at the beginning of the twentieth century. Two horses pull a sail reaper. They wear crocheted coverings over their ears and faces to protect them from flies. The people wear hats to ward off the sun.

A hard pull for the horses, leading the binder uphill.

The young boy takes the reins, the horses wait patiently while the sheaves of oats are loaded on to the wagon. In the background another wagon is led across the field for loading.

horses in the hayfield was limited to the carting. In the later period the power of horses became very important, for they pulled the mowing machine, hay maker and the horse rakes. A horse was also used to power the gearing to drive the elevator at the stack. If John Stewart Collis's horses had found harrowing tedious, then working the horse gear was the height of boredom. It was a job so few horses cared for that in some districts the one that would work the gear with equanimity was called the willing horse. It also was usually one of the oldest horses on the farm.

Working through the hot days of summer the horses could quickly get out of condition. At no time was this more likely than during the corn harvest. The days could be at their hottest then, just when the work was particularly busy, and in consequence the horses tended to tire easily. It was largely because of that, and the need for constant power, that cutting the corn was amongst the first tasks that farmers turned over to the tractor. The binders, which became the standard harvesting machine from the early twentieth century, were heavy, regularly requiring three horses, sometimes four. To keep the work in the harvest field going as late into the evening as possible—until 7 or 8 o'clock, and maybe later—on large farms, at least, a fresh team of horses took over the binders half way through the day.

Carting the corn to the stack was just as busy. It was often organised into teams of two wagons, each wagon drawn by two horses. A boy did the driving while the horseman directed the loading. Well-trained and experienced horses were put to these wagons, for they could be relied upon to walk steadily to the stooks waiting to be loaded, with the men safely on top of the half-loaded wagon. Leading the wagon back to the yard, the boy walked with the shaft horse, driving the trace horse with a plough line.

Harvest brought the horseman's year to its busy climax, and then following a short break the autumn ploughing of the stubbles began again.

Four horses ready to take the strain of the harvest wagon, on the South Downs near Lewes, 1928.

5

SHOEING AND SADDLERY

ON wet days there was likely to be a steady procession of horses arriving at the blacksmith's shop to be shod. They would start arriving by about 7 o'clock and would keep the smith busy until late in the afternoon, while all the horsemen waiting their turn spent the time in gossip with their neighbours. This was not unusual, since every farmer and foreman in the district, having seen that prospects for work in the fields were poor, had concluded that the horses might profitably pay a visit to the smith.

The blacksmith was one of the country craftsmen vital for the support of farming with horse power. Most of his work came from the farmers. The horses needed regular care and attention to their feet, and shoeing provided the steady, bread-and-butter trade for the smith. On top of that he had the repairs to the ploughs, harrows and other implements of the farm.

'No foot, no horse', ran an old adage, and attention to the feet was an essential part of the horseman's work. In particular he had to see that the horses were well shod. Shoes were worn to protect the horny part of the hoof (the wall) from being broken or worn down too much by contact with hard or stony ground. At the same time, however, the shoe also shielded the horn from its natural process of wear, thus allowing the horn to grow behind and over the shoe, and this had to be checked by changing the shoe and paring the hoof every so often.

Just how often this should be done was a somewhat open matter. The writers of texts on both agriculture and farriery all agreed that four or five weeks was the time it took for the hoof to become overgrown and therefore dictated the frequency at which the horse should be shod. John Burke went so far as to recommend three weeks as the interval for removing the shoes, there being, he believed, 'few more vicious species of economy than putting heavy and long-lasting shoes upon any horse.'[7] At the other extreme George Ewart Evans recounts the recollections of a blacksmith from Needham Market, in Suffolk, who said that farm horses used to come in once every three months, on average.[16] There may have been slight exaggeration here, but many farmers certainly did ignore all the advice and put off having their horses shod. This was particularly likely during the 1920s and 1930s, the time recalled by this last smith, when the hard times encouraged parsimony.

When the horse arrived to be shod the first thing the smith did was to examine the feet carefully. As a farrier, he was expert in the care of the horse's feet, and was able to assist the farmer with many of the problems of unsound feet to which horses were prone. In fact, he usually knew far more of the horse than its feet, coming to know its whole physical wellbeing and its temperament as well, sometimes better than the farmer who owned it. The smith, therefore, was likely to be a source of unofficial advice on the quality of the local horseflesh.

The smith started his examination with the fore feet. The horse was more likely to keep calm throughout the proceedings if it could see what was going on at the beginning. That was not foolproof, of course, and there were always horses that were too nervous, or simply uncooperative. Instead of being lightly tethered, or standing free, some horses had to be restrained more firmly, even perhaps fettered completely, and their feet shod while they were laid on their back.

Having made this initial examination, the smith would take a buffer and pair of pincers from his farrier's box, in which were kept all the tools reserved for shoeing—on no account to be used for any general smithing work. The buffer was for breaking off

There really were smiths who had chestnut trees outside their forges. The master farrier looks on as his assistant attends to the fore foot of the horse. The young horseman keeps by the horse's head to reassure the animal.

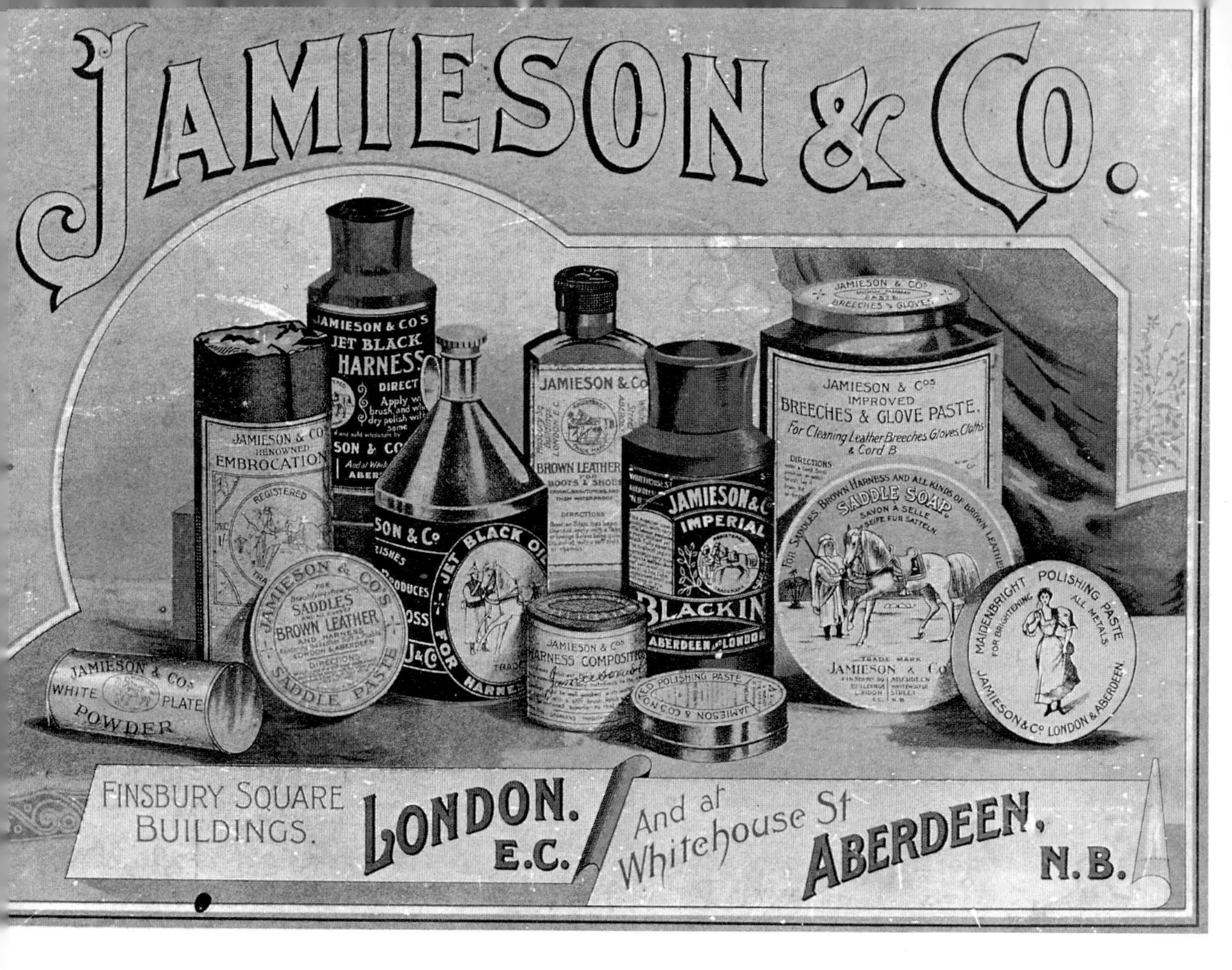

An advertising card that hung for many years in a saddler's shop in a Berkshire village. The card dates from the first years of the twentieth century and advertises the products of one of the leading suppliers of saddle soaps, harness blacking and embrocation that was rubbed into the limbs of horses as a preventative against rheumatism.

the heads of nails, and at its simplest might be a metal plate sharpened at one edge. More elaborate ones also existed, rather like a flattened letter H, with the cross-bar having the sharp edge. Having removed the nail heads, the pincers were then used to prise away the shoe from the hoof.

The feet were then cleaned and pared before being reshod. The excess growth of horn was pared using rasp, paring knife and the pincer-like hoof clippers. The sole of the foot was given a thorough clean, with care being taken not to damage it or the frog, the triangular shaped pad of fibrous material at the rear of the sole which acts as an elastic cushion against the ground.

New shoes could then be fitted, or the old ones put back if they were not too worn. New shoes were made to fit the horse exactly. Particularly for new colts having their first shoes fitted, they might be fashioned while the horse waited, a length of bar iron being cut off, heated in the furnace and moulded and beaten to shape. Some country smiths kept a stock of shoes made to the measurements of their regular customers' horses; others would simply have the basic horse shoe that had been prepared when business was quiet. By the late nineteenth century there were a number of firms supplying factory-made shoes, which could be bought by the ton. Such firms issued illustrated catalogues showing a profusion of shoes for every kind of horse, from pony to hunter to draught horse, and offered shoes for both work and shows. Most smiths in the country districts, however, were still making their own shoes, at least until the 1920s.

The shoes from stock, even those made to the measurements for a particular horse, would need some altering and adapting to get the perfect fit. They were, therefore, heated in the hearth and offered up to the hoof, which created a lot of smoke as the horn of the hoof burned. It looked, and smelled, worse than it was, for the horn is quite insensitive. The shoe was reshaped as necessary, and once the smith was satisfied he had a perfect fit the shoe could be nailed in.

The nails, too, were by tradition made at the smithy, to a particular design, countersunk so that the head did not stand proud of the shoe. There were different sizes for the different horses. Carthorses needed the biggest, about three inches long, whereas the shoes on a hunter required nails only about two inches long.

This was the shop in which the preceding sign hung. The calendar is for 1962; otherwise the scene seems little different from the 1920s.

During winter times of hard frost and snow, horses needed some extra grip on the slippery roads and tracks. For that they were brought along to the smith's for roughing, which involved either shoeing with rough shoes, or putting some temporary grips—frost nails, or sharps—on the ordinary shoes. A sudden frost could result in the smith being rushed off his feet for a day, as every horse in the neighbourhood arrived at his door to be roughed.

The blacksmith's work for the horses did not stop at shoeing them, for he made and repaired the iron work for swingletrees, and all the hooks and chains for the harness.

The saddler and harness maker was another craftsman upon whom the farmer depended. He made and mended all the harness for the farm and supplied almost all the accessories for managing horses—whips, buckles and brasses for the harness, pastes and polishes for both the leather and metal in the harness, and embrocations.

The harness maker was from time to time called upon to make good the farmer's own shortcomings, as a number of farmers skimped on the harness pastes and expected their harness to last for decades, despite heavy wear and neglect. It fell to the saddler to achieve this longevity.

Most of the saddler's work for the farmer was on repairs. Some country saddlers actually made no more than a handful of the larger items—collars and cart saddles—in all their working lives. This was especially likely in later decades, when saddlers in the villages were more inclined to turn to firms in the West Midlands that specialised in supplying all the accoutrements for horses and their keepers. From these firms could be bought ready-made collars, saddles, bridles and all the other straps for the harness; and also the various spare parts, such as saddle-trees and buckles, which might be needed for repairs. Buying stock offered many conveniences to the saddler. He was saved some of the more awkward tasks, such as attempting to measure precisely the horse's neck, which was never easy. Instead, a selection of collars would be tried on until one fitted well. This diminished the bespoke nature of harness making, but for many farmers convenience and cheapness weighed more heavily.

Whether making or repairing, the saddler's skills were the same. Indeed, they were, perhaps, put more to the test by some of the repairs, when the saddler was presented with some worn specimen of leather to which he was expected to impart new life. The need for thorough knowledge of working with leather, and of every detail of the manufacture and repair of harness, kept alive the system of long apprenticeships of seven years with little pay. It remained strong into the inter-war years, when business for the saddler began to fall away.

The saddler sews up a well-worn collar that has come for another repair.

Not all the saddler's work was done at his shop. Many of the large farmers preferred him to come out to the farm to overhaul the complete stock of harness, perhaps once a year. A suitable lull in farming activities would be chosen, a favourite being the few weeks before harvest. The saddler would charge a set rate for the day for this service. He came out early in the morning to set up shop in a barn, or, in fine summer weather, out in a yard. He had to bring all of his tools with him, and there were a great many of these; saddlery and harness making having more special tools than most crafts. There were the half-moon and quarter-moon knives for cutting the leather, along with paring knives and edge shavers for trimming all the different thicknesses. There was also a range of pricking wheels or straight pricking irons, for marking the lines for stitching. For basic sewing five stitches to the inch were used, whereas up to 16 stitches to the inch were used for delicate work. To match the prickers was a range of awls for punching the holes and needles (always blunt ended) for the stitching. Such a range was needed since one of the marks of the skilled saddler was his ability to produce stitching of exactly the right quality for each job.

Saddlery used a great variety of leathers. In the early decades of the nineteenth century, many country saddlers still dressed their own skins, especially if they were using horse hide (white leather). Some were still following another ancient practice, slaughtering the horses themselves, and in Suffolk at least, as George Ewart Evans found, saddlers were often also known as knackers.

For farm harness cowhide became the main material. The best part of the hide, the back, was used for making new harness. Long strips cut down the middle produced the reins, breechings and other straps. The thinner and weaker hide of the belly was used mainly for repairs and linings, with some of the better parts used as side pieces on the collar. Heavy cart harness was usually dyed black.

While blacksmiths and farriers commonly had shops in the villages, there were fewer saddlers and harness makers who did so. There were many more in the market towns, where they catered for all the variety of riding, driving and van harness, in addition to their farm work. Some of these saddlers had quite large businesses, employing more than twenty men as well as apprentices, into the early twentieth century.

REFERENCES AND SELECT BIBLIOGRAPHY

Journal of the Royal Agricultural Society of England is cited as *JRASE*.

1. *A Century of Agricultural Statistics: Great Britain 1866–1966*, HMSO (1966).
2. Baker, R., 'The Farming of Essex', *JRASE*, v (1844).
3. Baker, Thomas J. L., 'On the Draught of Single Cart Horses', *JRASE*, i (1840).
4. Barnett, Margaret, *British Food Policy in World War I* (1988).
5. Barrett, Hugh, *Early to Rise* (1967).
6. Biddell, Herman, et al., *Heavy Horses: Breeding and Management* (6th ed., 1919).
7. Burke, John, 'Breeding and Management of Horses on a Farm', *JRASE*, v (1844).
8. Burn, R. Scott, *Outlines of Modern Farming* (4th ed., 1878).
9. Carter, I. *Farmlife in North East Scotland 1840–1914* (1979).
10. Chivers, Keith, *The Shire Horse: a History of the Breed, the Society, and the Men* (1976).
11. Collins, E. J. T., 'The Farm Horse Economy of England and Wales in the Early Tractor Age 1900–40', in F. M. L. Thompson (ed.), *Horses in European Economic History: a preliminary canter* (1983).
12. Collis, John Stewart, *While Following the Plough* (1946).
13. Day, Herbert L., *When Horses were Supreme* (1985).
14. Dent, Anthony, *Cleveland Bay Horses* (1978).
15. Dewey, Peter E., *British Agriculture in the First World War* (1989).
16. Evans, George Ewart, *The Horse in the Furrow* (1960).
17. Evans, George Ewart, *Horse Power and Magic* (1979).
18. Fairfax-Blakeborough, J., *Yorkshire Village Life, Humour and Characters* (nd).
19. *Farmers' Dictionary*, (c. 1840).
20. *Farmers' Weekly*, 31 January 1936; 25 May 1950.
21. *Farm Horses*, Young Farmers' Club booklet No. 13.
22. Fawcus, Henry E., 'Horse Breeding in Yorkshire', *JRASE*, lxxii (1911).
23. Fream, W., *Elements of Agriculture* (1892).
24. Hasluck, Paul N., *Saddlery and Harness-Making* (1904, reprinted 1962).
25. Henderson, George, *Farmer's Progress* (1950).
26. Hogg, Garry, *Hammer and Tongs* (1964).
27. Hunting, William, *The Art of Horse Shoeing* (3rd ed., 1899).
28. Keegan, Terry, *The Heavy Horse: Its Harness and Decoration* (1973).
29. Kightly, Charles, *Country Voices: Life and Lore in Farm and Village* (1984).
30. Langdon, John, *Horses, Oxen, and Technological Innovation: the Use of Draught Animals in English Farming from 1066 to 1500* (1986).
31. Legard, George, 'Farming of the East Riding of Yorkshire', *JRASE*, ix (1848).
32. Little, Edward, 'Farming of Wiltshire', *JRASE*, v (1844).

33. Low, David, *The Elements of Practical Agriculture* (1834).
34. McConnell, Primrose, *The Diary of a Working Farmer* (1906).
35. MacDonald, W., 'On the Relative Profits to the Farmer from Horse, Cattle and Sheep Breeding, Rearing and Feeding in the United Kingdom', *JRASE*, 2nd series, xii (1876).
36. MacNeilage, Archibald, 'Feeding and Management of Work Horses', *Transactions of the Highland and Agricultural Society of Scotland*, 5th series, iv (1892).
37. Major, J. K., *Animal-Powered Engines* (1978).
38. Malden, W. J., 'The Economical Management of Manual and Horse Labour', *JRASE*, 3rd series, vii (1896).
39. Morton, J. C., 'The Cost of Horse Power', *JRASE*, xix (1858).
40. Palin, William, 'The Farming of Cheshire', *JRASE*, v (1844).
41. Pitt, W., *General View of the Agriculture of the County of Worcester* (1813).
42. *Practical Power Farming*, December 1956.
43. Pusey, Philip, 'On the Progress of Agricultural Knowledge during the past Eight Years', *JRASE*, x (1850).
44. Read, Clare Sewell, 'Recent Improvements in Norfolk Farming', *JRASE*, xix (1858).
45. Spooner, W. C., 'On the Management of Farm Horses', *JRASE*, ix (1848).
46. *Standard Cyclopedia of Modern Agriculture* (c. 1915).
47. Stephens, Henry, *The Book of the Farm* (1849).
48. Stirton, Thomas, 'Select Farms in the Counties of Leicester and Rutland', *JRASE*, 3rd series, vii (1896).
49. Street, A. G., *Country Days* (1933).
50. Street, A. G., *Feather Bedding* (1954).
51. Street, A. G., *Round the Year on the Farm* (1941).
52. Tanner, Henry, 'The Agriculture of Shropshire', *JRASE*, xix (1858).
53. Vancouver, Charles, *General View of the Agriculture of Devon* (1808).
54. Wrighton, J., and Newsham, J. C., *Agriculture, Theoretical and Practical* (1919).
55. Youatt, W., *The Horse* (1851).
56. Young, Arthur, *General View of the Agriculture of the County of Lincolnshire* (1813).
57. Young, Arthur, *General View of the Agriculture of Norfolk* (1804).
58. Young, Arthur, *General View of the Agriculture of Oxfordshire* (1813).
59. Young, Arthur, *General View of the Agriculture of the County of Suffolk* (1813).

A NOTE ON WEIGHTS AND MEASURES

Contemporary weights and measures have been retained throughout the text. Thus acres, bushels, pounds, feet and inches are regularly mentioned.

For those less familiar with these, here are some basic metric equivalents.

The bushel is a unit of measure, but by the late nineteenth century there were conventional standard equivalent weights. For oats, a bushel was 42 lb. A pound is approximately 454 grams, so a bushel of oats becomes approximately 19 kg.

A hundredweight is 112 lb, or 50.8 kg.

An acre is 0.4 hectares.

A mile is 1.6 kilometres, one foot is 0.3048 metres, an inch is 2.54 centimetres.

The height of horses is measured in hands, a hand being 4 inches (the expression 16.2 hands means 16 hands and 2 inches). The horse is measured when standing square, from the ground to the wither (shoulder).

INDEX

Page numbers in bold refer to illustrations.

Farming Press Books

Listed below are a number of the agricultural and veterinary books published by Farming Press. For more information or a free illustrated book list please contact:

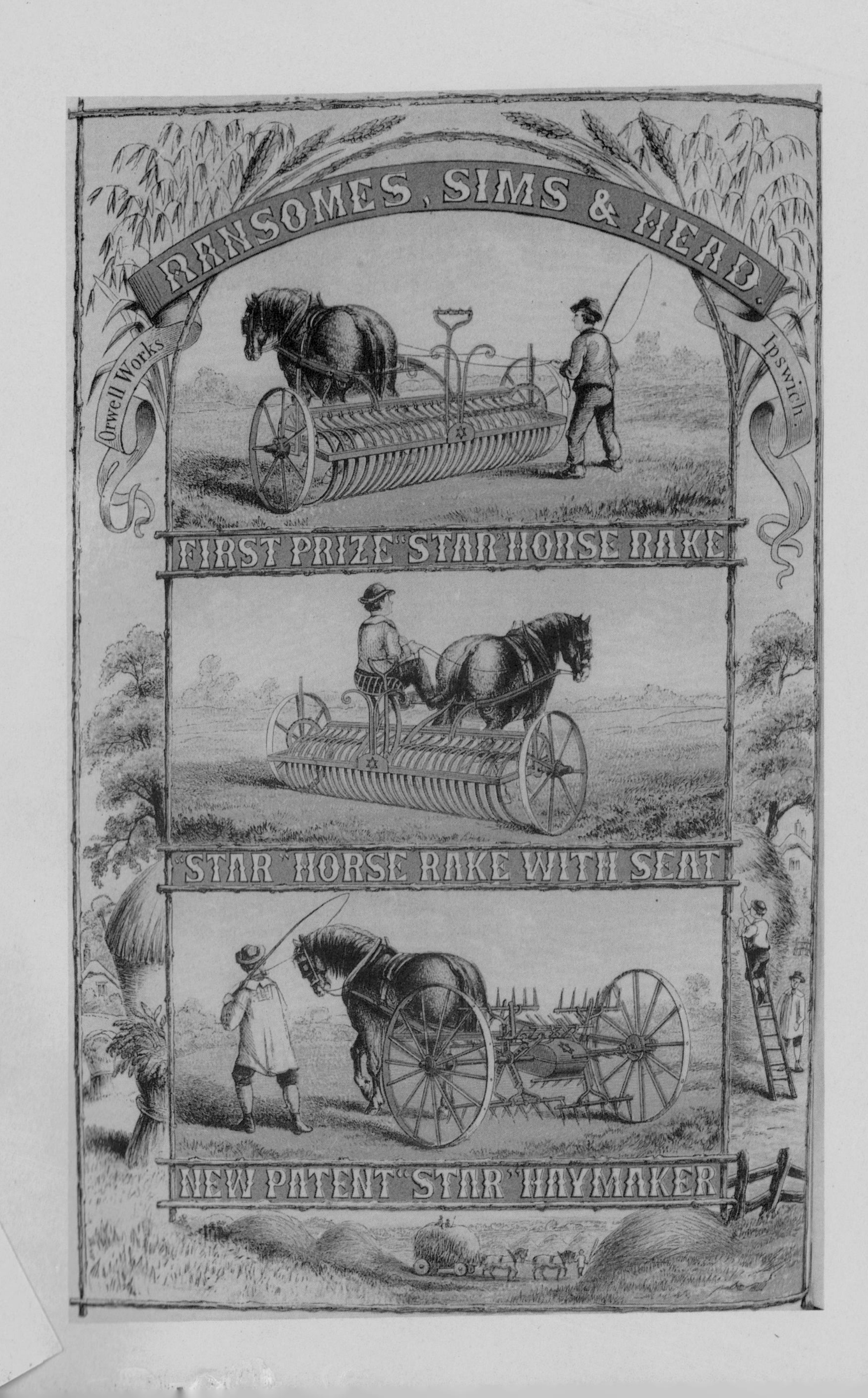

RANSOMES, SIMS & HEAD.
Orwell Works
Ipswich.
FIRST PRIZE "STAR" HORSE RAKE
"STAR" HORSE RAKE WITH SEAT
NEW PATENT "STAR" HAYMAKER